AA004051

# eTELEMED 2025

The Seventeenth International Conference on eHealth, Telemedicine, and Social Medicine

May 18 - 22, 2025

Nice, France

**eTELEMED 2025 Editors**

Svetlana Herasevich, Mayo Clinic - Rochester, USA

**Printed from e-media with permission by:**

Curran Associates, Inc.
57 Morehouse Lane
Red Hook, NY 12571

CURRAN ASSOCIATES INC.
proceedings
.com

**Some format issues inherent in the e-media version may also appear in this print version.**



International Academy, Research, and Industry Association (IARIA)
412 Derby Way
Wilmington, DE 19810

Phone: (408) 893-6407
Fax: (408) 527-6351

petre@iaria.org

**Additional copies of this publication are available from:**

Curran Associates, Inc.
57 Morehouse Lane
Red Hook, NY 12571 USA
Phone: 845-758-0400
Fax: 845-758-2633
Email: curran@proceedings.com
Web: www.proceedings.com

**Table of Contents**

# eTELEMED 2025

# Forward

The Seventeenth International Conference on eHealth, Telemedicine, and Social Medicine (eTELEMED 2025), held between May 18$^{th}$, 2025, and May 22$^{nd}$, 2025, in Nice, France, continued a series of events considering advances in techniques, services, and applications dedicated to a global approach of eHealth.

Development of wireless homecare, of special types of communications with patient data, videoconferencing and telepresence, and the progress in image processing and date protection increased the eHealth applications and services, and extended Internet-based patient coverage areas. Social and economic aspects, as well as the integration of classical systems with the telemedicine systems, are still challenging issues.

The event provided a forum where researchers were able to present recent research results and new research problems and directions related to them. The topics covered aspects from classical medicine and eHealth integration, systems and communication, devices, and applications.

We take here the opportunity to warmly thank all the members of the eTELEMED 2025 technical program committee, as well as all the reviewers. The creation of such a high-quality conference program would not have been possible without their involvement. We also kindly thank all the authors who dedicated much of their time and effort to contribute to eTELEMED 2025. We truly believe that, thanks to all these efforts, the final conference program consisted of top-quality contributions. We also thank the members of the eTELEMED 2025 organizing committee for their help in handling the logistics of this event.

We hope that eTELEMED 2025 was a successful international forum for the exchange of ideas and results between academia and industry for the promotion of progress in the field of telemedicine and social medicine.

**eTELEMED 2025 Chairs**

**eTELEMED 2025 Steering Committee**
Yoshitoshi Murata, Iwate Prefectural University, Japan
Les Sztandera, Thomas Jefferson University, USA
Vitaly Herasevich, Mayo Clinic, USA

**eTELEMED 2025 Publicity Chairs**
José Miguel Jiménez, Universitat Politècnica de València, Spain
Francisco Javier Díaz Blasco, Universitat Politècnica de València, Spain
Ali Ahmad, Universitat Politècnica de València, Spain
Sandra Viciano Tudela, Universitat Politècnica de València, Spain
Laura Garcia, Universidad Politécnica de Cartagena, Spain

## eTELEMED 2025
## Committee

### eTELEMED 2025 Steering Committee

Yoshitoshi Murata, Iwate Prefectural University, Japan
Les Sztandera, Thomas Jefferson University, USA
Vitaly Herasevich, Mayo Clinic, USA

### eTELEMED 2025 Publicity Chairs

José Miguel Jiménez, Universitat Politècnica de València, Spain
Francisco Javier Díaz Blasco, Universitat Politècnica de València, Spain
Ali Ahmad, Universitat Politècnica de València, Spain
Sandra Viciano Tudela, Universitat Politècnica de València, Spain
Laura Garcia, Universidad Politécnica de Cartagena, Spain

### eTELEMED 2025 Technical Program Committee

Shabbir Syed Abdul, Taipei Medical University, Taiwan
Don Adjeroh, West Virginia University, USA
Poovendran Alagarsundaram, Humetis Technologies Inc, Kingston, NJ, USA
Giovanni Albani, Istituto Auxologico Italiano - IRCCS, Italy
Basel Almourad, College of Technological Innovation, Dubai, UAE
Domingos Alves, Ribeirao Preto Medical School | University of Sao Paulo (USP), Brazil
Prima Oky Dicky Ardiansyah, Iwate Prefectural University, Japan
Antonio Augusto Goncalves, Estacio de Sá University, Brazil
Rafael Avila, Hospital Universitario Privado de Cordoba, Argentina
Mansoor Baig, King Faisal Specialist Hospital & Research Center, Riyadh, Saudi Arabia / ICIMTH, Greece
Panagiotis D. Bamidis, School of Medicine - Aristotle University of Thessaloniki, Greece
Oresti Baños, University of Granada, Spain
Ivana Bartoletti, Gemserv, UK
Azadeh Bashiri, Shiraz University of Medical Sciences, Iran
Hrvoje Belani, Ministry of Health - Directorate for e-Health, Zagreb, Croatia
Elisabetta Benevento, University of Pisa, Italy
Arriel Benis, HIT- Holon Institute of Technology, Israel
Sid-Ahmed Berrani, Ecole Nationale Polytechnique, Algiers, Algeria
Anna M. Bianchi, Politecnico di Milano, Italy
Vilmos Bilicki, University of Szeged, Hungary
Lucia Billeci, National Research Council of Italy | Institute of Clinical Physiology, Pisa, Italy
Antonis Billis, School of Medicine - Aristotle University of Thessaloniki, Greece
Tetiana Biloborodova, G.E. Pukhov Institute for Modelling in Energy Engineering, Ukraine
Sylvie Briand, World Health Organization, Switzerland
Taxiarchis Botsis, The Sidney Kimmel Comprehensive Cancer Center | Johns Hopkins University School of Medicine, USA
Eman Buhagiar, Middlesex University, Malta

Marco Buzzelli, University of Milano - Bicocca, Italy
Enrico Gianluca Caiani, Politecnico di Milano, Italy
Jessica Campbell, University of Central Florida / Optum, USA
Manuel Campos Martínez, University of Murcia, Spain
Nicola Carbonaro, University of Pisa, Italy
Ayan Chatterjee, The University of Agder, Grimstad, Norway
Darwyn Chern, Copa Health, Phoenix, USA
Bhargava Chinni, University of Rochester, USA
Mario Ciampi, National Research Council of Italy | Institute for High Performance Computing and Newtorking, Italy
James J. Cimino, Informatics Institute - University of Alabama at Birmingham, USA
Javier Civit, Cober SL, Spain / Gnomon Informatics, Greece
Daniel Condor Camara, Cayetano Heredia University, Peru
Massimo Conti, Università Politecnica delle Marche, Ancona, Italy
Sandra Costanzo, University of Calabria, Italy
Paul M. Cunningham, IST-Africa Institute, Ireland
Jacques Demongeot, Université Grenoble Alpes, France
Gayo Diallo, Univ. Bordeaux/ISPED, France
Linying (Lin) Dong, Ryerson University, Canada
Alexandre (Sasha) Douplik, Toronto Metropolitan University, Canada
Audrey DunnGalvin, University College Cork, Ireland
Duarte Duque, 2Ai | Polytechnic Institute of Cávado and Ave Barcelos, Portugal
Claudio Eccher, FBK Fondazione Bruno Kessler, Italy
Dina El Demellawy, CHEO Research Institute | University of Ottawa, Canada
Mohamed El Hafedh Abdi, Centre d'imagerie scintigraphique Blida, Algeria
Christo El Morr, York University, Canada
Radwa El Shawi, University of Tartu, Estonia
Manuel Filipe Santos, University of Minho, Portugal
Bruno Fionda, Fondazione Policlinico Universitario "A. Gemelli" IRCCS, Italy
Sebastian Fudickar, Universität Oldenburg, Germany
Niels F. Garmann-Johnsen, University of Agder, Norway
Anthony Gelibert, Carbon Bee, France
Wojciech Glinkowski, Polish Telemedicine Society / Center of Excellence "TeleOrto", Poland
Kuang Gong, Massachusetts General Hospital / Harvard Medical School, USA
Manuel González-Hidalgo, University of the Balearic Islands / Balearic Islands Health Research Institute (IdISBa), Spain
Conceição Granja, Norwegian Centre for e-health Research, University Hospital of North Norway, Norway
David Greenhalgh, University of Strathclyde, Glasgow, UK
Teresa Guarda, Universidad Estatal Peninsula Santa Elena - UPSE, Ecuador
Mohammad Hassanzadeh, Tarbiat Modares University, Iran
Vitaly Herasevich, Mayo Clinic, USA
Pilar Herrero, Universidad Politécnica de Madrid, Spain
Felix Holl, DigiHealth Institute - Neu-Ulm University of Applied Sciences / IBE - University of Munich, Germany
Delowar Hossain, BRAC University | United International University | IDCL Fiance Limited, Bangladesh
Ying-Feng Hsu, Osaka University, Japan
Yan Hu, Blekinge Institute of Technology, Sweden

Maryam Jafarpour, University of Tehran / Ministry of Health and Medical education, Iran
Tian Kang, Tempus Labs Inc., USA
Ashad Kabir, Charles Sturt University, Australia
Haralampos Karanikas, University of Thessaly, Greece
Martijn Kiers, University of Applied Science FH JOANNEUM, Austria
Toralf Kirsten, University of Applied Sciences Mittweida, Germany
Monika Knudsen Gullslett, Nasjonalt Senter for e-Helseforskning, Sweden
Evdokimos Konstantinidis, Aristotle University of Thessaloniki, Greece / Nively, Nice, France
Stathis Th. Konstantinidis, School of Health Sciences | University of Nottingham, UK
Valencia Koomson, Tufts University, USA
Frank Kramer, Faculty of Medicine/University of Augsburg, Germany
Vinay Kumar, Thapar University, Patiala,India
Mouhamadou Lamine Ba, Ecole Supérieure Polytechnique | Université Cheikh Anta Diop, Dakar, Sénégal
Daniele Landi, University of Bergamo, Italy
Ettore Lanzarone, University of Bergamo, Italy
Edward R. Laskowski, Mayo Clinic, USA
Carla V. Leite, University of Aveiro, Portugal
Ove Lintvedt, Norwegian Centre for E-health Research / Nord University, Norway
Siru Liu, University of Utah, USA
Tatjana Lončar-Turukalo, University of Novi Sad, Serbia
Ana Rita Londral, Nova University of Lisbon, Portugal
Guillermo Lopez Campos, Wellcome-Wolfson Institute for Experimental Medicine | Queen's University Belfast, UK
Ljerka Luic, University North, Croatia
Gang Luo, University of Washington, USA
Rafael Maestre Ferriz, CETEM, Spain
Flora Malamateniou, University of Piraeus, Greece
Sadouanouan Malo, University Nazi Boni, Burkina Faso
Luis Marco-Ruiz, Norwegian Centre for E-health Research | University Hospital of North Norway, Tromsø, Norway /Peter L. Reichertz Institute for Medical Informatics of TU Braunschweig |Hannover Medical School, Germany
Giancarlo Mauri, University of Milano-Bicocca, Italy
Enkeleint-Aggelos Mechili, University of Vlora, Albania / University of Crete, Greece
Julio César Mello Román, Universidad Nacional de Asunción, Paraguay
Alessandro Mengarelli, Università Politecnica delle Marche, Ancona, Italy
Robert Mischak, Graz University of Applied Sciences, Austria
Sandra Mitrovic, IDSIA - USI/SUPSI (Dalle Molle Institute for Artificial Intelligence), Switzerland
Layal Mohtar, American University of Beirut Medical Center, Lebanon
António H. J. Moreira, 2Ai - Polytechnic Institute of Cávado and Ave, Barcelos, Portugal
Fernando Moreira, Universidade Portucalense, Portugal
Yoshitoshi Murata, Iwate Prefectural University, Japan
Sahiti Myneni, The University of Texas | School of Biomedical Informatics, USA
Paolo Napoletano, University of Milan-Bicocca, Italy
Yuriy L. Orlov, Russian Academy of Sciences | The Digital Health Institute I.M. Sechenov, Russia
Nuria Ortigosa, Universitat Politecnica de Valencia, Spain
Anna Pastusiak, StethoMe® / Adam Mickiewicz University, Poznan, Poland
Hugo Peixoto, Algoritmi Research Center | University of Minho, Portugal
Vitor Pinheiro de Almeida, Pontifícia Universidade Católica do Rio de Janeiro (PUC-Rio), Brazil

Ivan Miguel Pires, Instituto de Telecomunicações | Universidade da Beira Interior, Covilhã, Portugal
Prasad Ponnapalli, Manchester Metropolitan University, UK
Filipe Portela, University of Minho, Portugal
Sandhya Prabhakaran, Moffitt Cancer Center, Tampa, USA
Rüdiger Pryss, University of Würzburg, Germany
Ilir Qose, Aicare Srl, Italy
Taoufik Rachad, University of Mohammed V, Rabat, Morocco
M. Sohel Rahman, Bangladesh University of Engineering and Technology, Bangladesh
Gurprit K. Randhawa, First Nations Health Authority / University of Victoria / McMaster University, Canada
Sónia Rolland Sobral, Universidade Portucalense, Portugal
Carlos Rompante Cunha, CeDRI & UNIAG & Polytechnic Institute of Bragança, Portugal
Juha Röning, University of Oulu, Finland
Hayri Sever, Cankaya University, Turkey
Gro-Hilde Severinsen, Norwegian centre for e-health research, Norway
Zubair Shah, Hamad Bin Khalifa University (HBKU), Qatar
Rosa Sicilia, University Campus Bio-Medico of Rome, Italy
Line Silsand, Norwegian Centre for E-health Research, Norway
Åsa Smedberg, Stockholm University, Sweden
Sweta Sneha, Kennesaw State University, USA
Alessandro Stefanini, University of Pisa, Italy
Vasile Stoicu-Tvadar, University Politehnica Timisoara, Romania
Muhammad Sulaiman, UiT The Arctic University of Norway, Norway
Kenji Suzuki, Tokyo Institute of Technology, Japan
Alessandro Tognetti, University of Pisa, Italy
Alessandro Tonacci, Institute of Clinical Physiology | National Research Council of Italy (IFC-CNR), Pisa, Italy
Niruwan Turnbull, Mahasarakham University, Thailand
Gary Ushaw, Newcastle University, UK
Aristides Vagelatos, CTI&P, Athens, Greece
José Luis Vázquez Noguera, Universidad Nacional de Asunción, Paraguay
Laura Vera Righi, National Cancer Institute, Uruguay
Henrique Vicente, University of Évora, Portugal
Dongwen Wang, Arizona State University, USA
Utoomporn Wongsin, Foundation for Research Institute on Social Protection and Health, Thailand
Takashi Yamauchi, Texas A&M University, USA
Ping Yu, University of Wollongong, Australia
Zhongming Zhao, University of Texas Health Science Center at Houston, USA
Huiru (Jane) Zheng, Ulster University, UK
Stelios Zimeras, University of the Aegean, Greece
Evi Zouganeli, OsloMet - Oslo Metropolitan University, Norway
Emmanouil A. Zoulias, School of Health Sciences - National and Kapodistrian University of Athens, Greece

**Copyright Information**

# Virtual Intercorporeality: Using the Body Scan Meditation to Enhance Interoceptive Awareness, Therapeutic Alliance and Presence in TeleMental Health

Eileen Wurst
California Institute of Integral Studies
San Francisco, CA
e-mail: ewurst@mymail.ciis.edu

*Abstract*—**Having to transition to TeleMental health during the COVID-19 pandemic, psychotherapists were challenged to continue establishing therapeutic connections in their sessions within this new human computer interaction. In the in-person psychotherapy exchange, the clinician and client form a mutual connection that helps them to feel a sense of therapeutic presence and alliance together. This is known in somatic psychology as intercorporeality, a critical component of the therapeutic relationship that is suspected to be less accessible through digital environments. Motivated by the scarcity of existing literature correlating the relationships between TeleMental health, human-to-human interaction with technology, intercorporeality, therapeutic presence, working alliance, and interoceptive awareness, this paper examines relevant empirical studies and presents a design for a mixed-methods study to assist in advancing understanding of these constructs. The explicit aim is to explore how a body scan meditation effects interoceptive awareness, working alliance and sense of therapeutic presence for both therapist and client within the context of the human-to-human interaction with technology within a TeleMental health session.**

***Keywords-human-to-human interaction with technology; somatic psychology; telemental health; body scan meditation.***

## I. INTRODUCTION

The (COVID-19) Corona Virus Disease 2019 pandemic necessitated a rapid transition from in-person psychotherapy to digital formats, reshaping therapeutic practices and interactions within virtual environments. Psychotherapists accustomed to physical presence and embodied connections with their clients faced challenges as they adapted to TeleMental Health (TMH), which employs video and other digital tools to deliver therapy remotely [1][2].

Despite TMH's benefits in accessibility and safety, its virtual nature often reduces the depth of interpersonal connections, impacting a core component of somatic thera called intercorporeality or the mutual, embodied awareness between therapists and their client [3][4]. One promising approach to enhancing intercorporeality in TMH is the integration of Body Scan Meditation (BSM). Rooted in ancient mindfulness practices, BSM encourages individuals to progressively focus on different areas of the body, fostering awareness of physical sensations without judgment [5]. Preliminary research suggests that BSM can improve Interoceptive Awareness (IA), reduce anxiety, and enhance overall presence by grounding participants in their sense of body and body type [6][7]. This practice may thus offer a way to recreate elements of embodied connection even within the constraints of a virtual setting. The proposed study in this paper was inspired and influenced by a previous phenomenological pilot study in which we investigated the experience of the intercorporeality within the TMH session on a clinician's sense of their own embodiment as well as that of their clients [8]. The following study aims to investigate whether implementing BSM within TMH sessions can bolster interoceptive awareness and foster a stronger sense of Therapeutic Presence (TP) and Working Alliance (WA) between therapists and clients. By focusing on these aspects, this research seeks to contribute to the evolving landscape of somatic virtual therapy, highlighting innovative methods for cultivating meaningful therapeutic connections within the digital context of TMH.

The rest of the paper is structured as follows. Section II discusses the relevant literature of the variables and constructs involved in this study. Section III provides an overview of the methodology of the research. Section IV reports on the current status of the investigation. Section V discusses the limitations imposed by self-report measures, small study sample, technological disruptions, and short duration. Section V presents the conclusion and considers future investigations and implications of the study.

## II. LITERATURE REVIEW

### A. *TMH and the Challenges of Embodied Connection*

There was a marked increase in the use of TMH leading into the pandemic, with a 22.3% rise in utilization reported from 2019 to 2022 [9]. The Utilization Index presented in Figure 1 sets 2019 as the baseline year (indexed at 100). TMH usage increased markedly in 2020, reaching an index value of 115, which reflects a 15% rise compared to 2019. Utilization peaked in 2021 at approximately 125, indicating a 25% increase over baseline. Although there was a slight decline in 2022, the index remained above 2019 levels, highlighting sustained use of TMH services beyond the pandemic surge. Since TMH continues to be utilized within the mental health field and will evolve with the technology it is part of, the embodied connection between therapist and client will remain an integral part of the therapeutic connection. BSM is explored in order to assist facilitation of the components of intercorporeality.

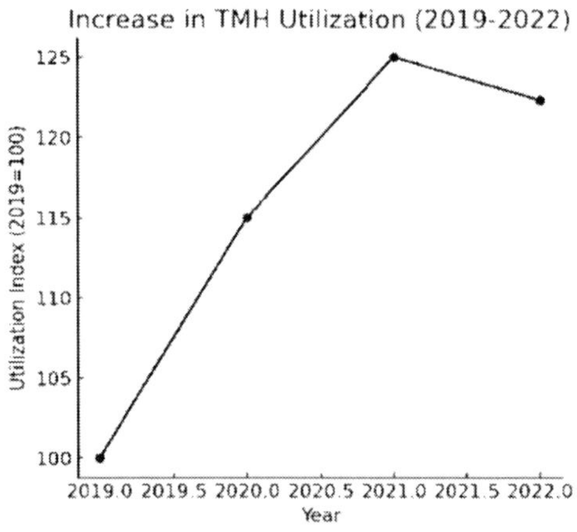


Figure 1. TMH Utilization 2019-2022 [9].

*B. BSM to Facilitate Therapeutic IA*

IA, or the conscious perception of internal bodily sensations, is closely linked to emotional regulation and TP. Techniques such as BSM that enhance IA have shown promise in fostering mindfulness and reducing anxiety [6][10]. BSM can help individuals connect to Increase in their bodily sensations, increase their sense of TP, and support emotional regulation [11][12][13].

IA has been investigated with in-person body- oriented therapies, focusing on its role in emotion regulation and self-awareness [14]. This research has suggested that IA therapies that incorporated BSM improved clients' ability to recognize and interpret physical sensations, resulting in better emotional processing

This paper refers to Human-to-Human Interaction with Technology (HHIT) as the digitally mediated communication between therapists and clients using platforms such as video conferencing [15]. However, there is limited research on somatic and mindfulness practices such as BSM application in HHIT environments like TMH, but preliminary findings from one investigation indicate that virtual BSM may enhance client and therapist attunement by fostering an embodied sense of TP, even in the absence of physical co-presence [3]. Additionally, virtual BSM has shown potential for reducing stress and anxiety, creating a more relaxed and connected therapeutic environment that facilitates the reinforcement of the WA between therapist and client [16].

*C. Role of TP within the HHIT of TMH*

In TMH, the concept of TP—the therapist's and client's awareness and engagement in the shared therapeutic space—is crucial for building trust and facilitating effective communication. Intercorporeality plays a crucial role in therapeutic effectiveness, facilitating deeper relational engagement and a sense of shared TP. In face-to-face therapy, this embodied connection is supported by nonverbal cues, such as body language and synchronized breathing, which foster client and therapist attunement. However, virtual platforms such as the HHIT of TMH limit these interactions, often resulting in a perceived detachment that can impact therapeutic outcomes [17][18]. Figure 2 depicts these deficits. Some research elucidated strengthening presence and IA may enhance the TMH experience [19][20].

Studies on presence in HHIT settings also highlight the importance of audio-visual cues, with larger screens and clearer visual input enhancing perceived presence and connection [21]. However, therapists often need to modify their behavior, such as using more demonstrative body language and ensuring a distraction-free environment, to foster presence in TMH [22].

Embodied practices, such as BSM, may enhance TP and intercorporeality in TMH by encouraging both therapists and their clients to tune into their bodily sensations, creating a more grounded and connected therapeutic interaction [23]. By integrating IA and mindfulness practices, TMH can potentially create conditions that approximate the embodied presence typically achieved in in-person settings.

*D. Role of Therapeutic Bond and WA in TMH*

Whether in-person or in TMH, the therapeutic bond is a foundational element of the WA, which also includes agreement on goals and tasks between client and therapist. The bond refers to the emotional connection and trust that develops, facilitating open communication and collaboration [24]. This bond is crucial for creating a safe, empathetic environment where clients feel understood and supported, which enhances the overall therapeutic WA. The advantages of a well-established psychotherapeutic relationship create a robust WA, characterized by trust, collaboration, and mutual understanding, establishing a secure foundation for the introduction of innovative therapeutic strategies such as BSM [25][26].

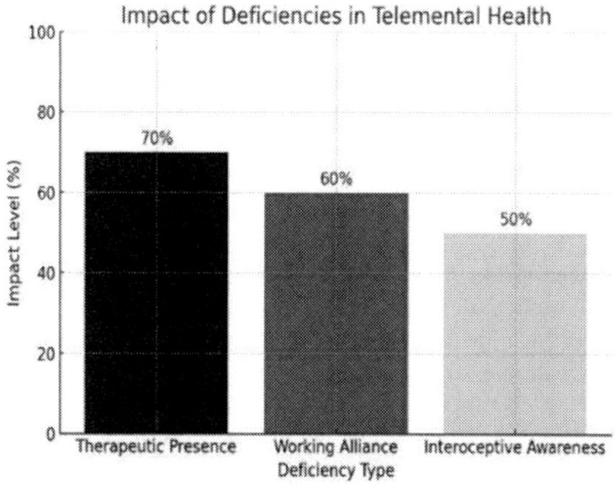


Figure 2. Estimated Impact of Deficiencies in TeleMental Health [17][18].

## III. Proposed Methodology

This study aims to examine the potential benefits of integrating a virtual BSM into TMH sessions, with a focus on enhancing IA, WA, TP, and thus, intercorporeality between therapists and clients during their HCI interaction. Specifically, the study investigates whether practicing BSM at the beginning of TMH sessions can foster a deeper therapeutic connection and improve clients' and therapists' experience across all the discussed constructs within the HHIT environment.

### A. *Study Design*

This study will employ mixed-methods research design, combining self-reports of all participants via three quantitative assessments and one qualitative questionnaire. A mixed-methods approach is well-suited for evaluating measurable changes in IA, WA and TP and gaining deeper insights into participants' personal experiences with BSM in a virtual setting [11]. Quantitative data will provide statistical measures of IA, WA and TP, while qualitative data will capture the nuanced experiences and reflections of all participants, offering a comprehensive understanding of BSM's effects on virtual therapy.

### B. *Participants*

A total of twenty-five participants will be recruited for this study: 20 clients and 5 therapists. Each therapist-participant will conduct a 5-minute BSM over TMH each week for six weeks. Eligible client-participants must be 18+, speak English fluently and engaged in therapy with their therapist-participant for at least 4 previous sessions. Participants will be excluded if they have prior experience with mindfulness practices or symptoms or conditions affecting their IA. These include autistic spectrum disorder, eating disorders or chronic pain diagnoses. These conditions have demonstrated difficulty in IA perception. This exclusion criteria are to ensure consistent IA baselines across participants [27][28].

### C. *Procedures*

- Therapist-participants will participate in a three-hour online workshop led by the researcher. The training will cover somatic psychology, applications in TMH and BSM principles, and a standardized BSM script for use in TMH sessions.
- Client-participants will be recommended by therapist-participants and will be screened and receive informed consent.
- At the beginning of each weekly TMH session, therapist-participants will lead a 5-minute BSM following the provided script. After the 1st, 3rd, and 6th sessions, all participants will complete the following measures:

### D. *Measures*

Quantitative:

- Scale of Body Connection (SBC): This instrument will measure changes in all participants' interoceptive awareness and overall body connection over the study period [29].
- The Working Alliance Inventory (WAI): Client-participants and therapist-participants will use their relevant version (WAI-C and WAI-T) to examine collaborative relationships from each other's perspectives [30].
- Therapeutic Presence Inventory (TPI): Client-participants will use the TPI-C to rate their perception of their therapist's presence, while therapist-participants will use the TPI-T to assess their own sense of presence and attunement during sessions [31].

Qualitative:

- Following the final session, a qualitative questionnaire with 4 open-ended questions will be conducted with both client-participants and therapist-participants. These interviews will explore both participants' experiences of IA, TP and perceived connection during TMH sessions with BSM, using as a guide Haley's open-ended questions [32].

### E. *Data Analysis*

Quantitative Analysis: Quantitative data will be analyzed using paired t-tests to compare mean pre- and post-intervention scores on the SBC, TPI, and WAI, and repeated measures Multivariate Analysis of Covariance (MANCOVA) to assess any interaction effects over time. This analysis will determine whether BSM produces significant improvements in IA, TP, and WA across the study period.

Qualitative Analysis: Questionnaire transcripts will be analyzed thematically following Braun and Clarke's methodology to identify common themes and insights [8]. This analysis will allow for a detailed exploration of participants' experiences with BSM in TMH, particularly in terms of intercorporeal connection and TP.

### F. *Treatment of Data*

Quantitative data will be handled in such a way that the criteria of academic ethics will be followed: precision, confidentiality, and the opportunity to reproduce the findings [3]. Initial cleaning procedures will include removing missing or inconsistent responses. Imputation will be performed only in very specific cases which would not compromise data integrity.

An explanatory sequential design will be employed to embed both quantitative and qualitative data together, thus enabling comprehensive analysis required regarding the effects of BSM within TMH. Baseline trends and quantified measures from quantitative research will help to put a real face into place with qualitative narratives of people involved. Triangulation will allow validation of results with enriching interpretation through cross-referencing numerical patterns and trends from quantitative data with those patterns identified within the lived experience reported from the qualitative data collection process.

## IV. CURRENT STATUS

Currently, this study is in the second month of recruitment. During the first month, multiple postings were made to over 30 social media groups for therapists across the

California Inst. of Integral Studies community, on Facebook, LinkedIn, listservs for therapists as well as the American Psychological and Counseling Associations. Thus far, 55 respondents have contacted this researcher. Unfortunately, most of the respondents (50) did not meet the criteria or were AI bots. Of the 5 who met the criteria, 2 have had to remove themselves from the study due to other time commitments and clients unable to participate. Currently we have 4 new therapists who have completed their paperwork and are encouraging their clients to return the paperwork so they can be screened. 2 of these therapists have completed the online preparatory training and will soon be implementing BSM in their sessions. Another round of online recruitment is being planned to recruit more participants in the event of attrition.

## V. Limitations

While this study aims to provide valuable insights into the effects of BSM on IA, TP, WA and intercorporeality in TMH, several limitations should be acknowledged:

- Sample Size and Diversity: The study's relatively small sample size (25 participants) may limit the generalizability of its findings. Additionally, the inclusion criteria, which exclude individuals with certain conditions (e.g., autism spectrum and eating disorders and chronic pain), could limit the applicability of the results to broader populations. Future research should aim to include a larger, more diverse sample to strengthen external validity.
- Self-Report Bias: The study relies on self-reported data through surveys and interviews, which may introduce biases such as social desirability or recall bias. Participants might report positive outcomes due to perceived expectations rather than actual changes in IA, WA or TP.
- Employing more objective measures, such as physiological indicators of IA (e.g., heart rate variability) in future studies, could provide additional validation.
- Duration and Consistency of Practice: This study is conducted over a six-week period, with a brief (5-minute) BSM exercise at the beginning of each session. While this design is practical, a longer duration or more frequent BSM practice could yield stronger effects and more sustained improvements in IA, WA, and TP. Further studies might investigate different durations or intensities of BSM to explore whether more frequent or prolonged practices offer greater benefits.
- Technology and Environment Variability: TMH sessions vary widely in participants' technology (e.g., screen size, internet stability) and environments (e.g., privacy, ambient noise), influencing the study's results. For instance, a larger screen may enhance a sense of TP, yet, interruptions in connectivity may disrupt engagement. The choice of platform for therapist-client interaction is not within our control, as participants are conducting sessions using their existing teletherapy tools. In order to broaden the potential candidate pool for recruitment, we have adopted a platform-agnostic approach. The focus of this study is on the use of BSM as a mechanism to enhance IA and TP within the WA of the TMH session—regardless of platform differences such as screen size or interface. In future iterations of the study, with a larger sample size and increased research funding, standardizing the platform across participants may be explored to control for this variable more tightly.

## VI. Conclusion

The shift to TMH has presented unique challenges and opportunities in maintaining presence and connection, particularly with the diminished intercorporeal exchange that in-person settings provide. In TMH, where physical presence is limited, alternative strategies are needed to foster the embodied awareness and attunement that support a meaningful therapeutic bond. This study proposes that BSM by enhancing IA, WA and TP, may prove to be a valuable tool for addressing these gaps and deepening the therapeutic intercorporeality in virtual environments. Practices that cultivate IA—such as BSM—are associated with greater self-awareness, emotional regulation, and a sense of grounded presence [6][7].

This study aims to contribute to the growing field of the integration of somatic practices within the HHIT of therapeutic practices such as TMH. Through a mixed-methods approach, it will offer both quantitative insights into the measurable benefits of BSM and qualitative perspectives on participants' lived experiences. If effective, BSM could serve as a foundational practice for improving the HHIT of virtual therapeutic relationships, informing future research and guiding practitioners in adapting somatic techniques to digital formats.

Likewise, by fostering a shared, embodied experience, BSM could help therapists overcome some of the interpersonal challenges presented by HHIT formats, providing a practical tool for strengthening connection and empathy with clients in remote sessions.

Additionally, these findings may encourage the development of training programs that introduce TMH practitioners to somatic techniques like BSM, equipping them to better support their clients' mental health in virtual environments. Future studies might expand on this research by examining the effects of BSM in diverse client populations or exploring other somatic practices that enhance WA, TP and IA in TMH.

## References


[1] E. Galvin et al., "Implementation of telemedicine consultations for people with mental health conditions in the community: a protocol for a systematic review," HRB Open Res., vol.4, 2021. [Online]. Available: https://doi.org/10.12688/hrbopenres.13435.3

[2] E. Whaibeh, H. Mahmoud, and H. Naal, "Telemental health in the context of a pandemic: The COVID-19 experience," Curr. Treat. Options Psychiatry, vol. 7, pp. 198-202, 2020.

[3] T. Fuchs, "Intercorporeality and interaffectivity," Phenomenol. Cogn. Sci., vol. 16, no. 4, pp. 491-503, 2017. [Online]. Available: https://doi.org/10.1007/s11097-016-9469-2.

[4] B. T. Villalobos, A. R. Dueweke, R. Orengo-Aguayo, and R. W. Stewart, "Patient perceptions of trauma-focused telemental health services using the Telehealth Satisfaction Questionnaire (TSQ)," Psychol. Serv., vol. 20, no. 1, pp. 107-121, 2023. [Online]. Available: https://doi.org/10.1037/ser0000605.

[5] B. Anālayo, "Buddhist antecedents to the body scan meditation," Mindfulness, vol. 11, no. 1, pp. 194-202, 2020. [Online]. Available: https://doi.org/10.1007/s12671-019-01259-8

[6] M. Dambrun et al., "Body scan meditation and its effects on interoceptive awareness: An integrative review," Mindfulness, vol. 13, no. 1, pp. 123-136, 2022. [Online]. Available: https://doi.org/10.1007/s12671-021-01647-4.

[7] K. Werner et al., "Interoceptive awareness, body scan meditation, and emotional regulation," Clin. Psychol. Rev., vol. 34, no. 6, pp. 775-787, 2014. [Online]. Available: https://doi.org/10.1016/j.cpr.2014.10.003.

[8] E. Wurst, "A qualitative pilot study on therapeutic embodiment in telemental health," in Proc. 15th Int. Conf. eHealth, Telemedicine, Social Med. (eTELEMED), Venice, Italy, 2023.

[9] R. McBain, J. Bauer, S. Victor, and T. Rhodes, "The rise of telemental health: A comprehensive review," Mental Health Services Review, vol. 19, no. 4, pp. 267-278, 2023.

[10] M. Schlief, D. Halliday, J. Stewart, K. Brindley, and N. Hunter, "Synthesis of the evidence on what works for whom in telemental health: Rapid realist review," Interactive Journal of Medical Research, vol. 11, no. 2, p. e38239, 2022.

[11] B. Harris, "Queer as a Bell: Music and the psychotherapeutic relationship," Voices: A World Forum for Music Therapy, vol. 19, no. 3, 2019. [Online]. Available: https://doi.org/10.15845/voices.v19i3.2674.

[12] C. Price and H. Weng, "Facilitating adaptive emotion processing and somatic reappraisal via sustained mindful interoceptive attention," Frontiers in Psychology, vol. 12, p. 578827, 2021.

[13] K. Werner, E. Jazaieri, W. Goldin, J. Ziv, H. Heimberg, and J. Gross, "Interoceptive awareness, body scan meditation, and emotional regulation," Clinical Psychology Review, vol. 34, no. 6, pp. 775-787, 2014. [Online]. Available: https://doi.org/10.1016/j.cpr.2014.10.003.

[14] C. Price, K. Krycka, T. Breitenbucher, and N. Brown, "Perceived helpfulness and unfolding processes in body-oriented therapy practice," The Indo-Pacific Journal of Phenomenology, vol. 11, no. 2, pp. 1-15, 2011.

[15] Z. E. Imel, D. D. Caperton, M. Tanana, and D. C. Atkins, "Technology-enhanced human interaction in psychotherapy," Journal of Counseling Psychology, vol. 64, no. 4, pp. 385-393, Jul. 2017, doi: 10.1037/cou0000213.

[16] L. R. Kogan and C. Bussolari, "Exploring the potential impact of a virtual body scan meditation exercise conducted with pet dogs on recipients and facilitators," Frontiers in Psychology, vol. 12, p. 698075, 2021.

[17] M. Ackerman, E. Greenwald, P. Noulas, and C. Ahn, "Patient satisfaction with and use of telemental health services in the perinatal period: A survey study," Psychiatric Quarterly, vol. 92, no. 3, pp. 925-933, 2021. [Online]. Available: https://doi.org/10.1007/s11126-020-09874-8.

[18] S. L. Connolly, C. J. Miller, J. A. Lindsay, and M. S. Bauer, "A systematic review of providers' attitudes toward telemental health via videoconferencing," Clinical Psychology: Science and Practice, vol. 27, no. 2, p. e12311, 2020. [Online]. Available: https://doi.org/10.1111/cpsp.12311.

[19] M. Ahlquist and B. Yarns, "Enhancing embodied connections in telemental health: Challenges and future directions," J. Somatic Psychol., vol. 15, no. 2, pp. 113-129, 2022. doi: 10.1080/02668734.2022.2037691.

[20] D. Schultchen et al., "Mindfulness and body awareness in somatic psychotherapy: An empirical analysis," Front. Psychol., vol. 10, p. 2358, 2019.

[21] A. M. Dickey and M. M. Wasko, "Digital disparities in patient adoption of telemedicine: A qualitative analysis of the patient experience," Int. J. Healthcare Inf. Syst. Inform., vol. 18, no. 1, pp. 1-21, 2023. doi: 10.4018/IJHISI.318043.

[22] S. P. Vidolov, "Virtual collaboration as co-enacting intercorporeality," Eur. J. Inf. Syst., vol. 33, no. 2, pp. 244-266, 2022.

[23] J. F. Tantia, The Embodied Therapist: Somatic Psychology and Clinical Practice. New York, NY, USA: Routledge, 2020.

[24] R. B. Ardito and D. Rabellino, "Therapeutic alliance and outcome of psychotherapy: Historical excursus, measurements, and prospects for research," Front. Psychol., vol. 2, p. 270, 2011. doi: 10.3389/fpsyg.2011.00270.

[25] B. Blasko, G. Serran, and J. Abracen, "The role of the therapeutic alliance in offender therapy," in Springer eBooks, 2018, pp. 87-108. doi: 10.1007/978-3-030-01030-0_5.

[26] B. Harris, "Queer as a Bell: Music and the psychotherapeutic relationship," Voices World Forum Music Ther., vol. 19, no. 3, 2019. doi: 10.15845/voices.v19i3.2674.

[27] C. Costin, "The centrality of presence and the therapeutic relationship in eating disorders," in Trauma-Informed Approaches to Eating Disorders, A. J. Seubert and P. Virdi, Eds. New York, NY, USA: Springer Publishing Company, LLC, 2019, pp. 45-56.

[28] A. Fotopoulou and M. Tsakiris, "Mentalizing homeostasis: The social origins of interoceptive inference," Neuropsychoanalysis, vol. 19, no. 1, pp. 3-28, 2017. doi: 10.1080/15294145.2017.1294031.

[29] C. J. Price et al., "The scale of body connection: Validation of a measure of embodiment," Mindfulness, vol. 8, no. 3, pp. 441-452, 2017. doi: 10.3389/fpsyg.2021.578827.

[30] E. S. Bordin, "The generalizability of the psychoanalytic concept of the working alliance," Psychother. Theory Res. Pract., vol. 16, no. 3, pp. 252-260, 1979. doi: 10.1037/h0085885.

[31] S. M. Geller and L. S. Greenberg, TP: A Mindful Approach to Effective Therapy. Washington, DC, USA: American Psychological Association, 2012.

[32] American Psychological Association, Publication Manual of the American Psychological Association, 7th ed. Washington, DC, USA: American Psychological Association, 2020.

# Size Matters: E-Health Implementation Challenges Across Norwegian Municipalities

Asbjørn Johansen Fagerlund
Norwegian Centre for E-health Research
Tromsø, Norway
email: Asbjorn.johansen.fagerlund@ehealthresearch.no

Rune Pedersen
Norwegian Centre for E-health Research
Tromsø, Norway
email: Rune.pedersen@ehealthresearch.no

Kristian Malm-Nicolaisen
Norwegian Centre for E-health Research
Tromsø, Norway
email: Kristian.nicolaisen@ehealthresearch.no

*Abstract*— **This study explores the size-dependent challenges and strategies in implementing e-health systems across Norwegian municipalities. Through a qualitative analysis of semi-structured interviews, the research identifies distinct barriers and approaches among small, medium, and large municipalities. Small municipalities face operational challenges and resource constraints, medium municipalities struggle with system integration and interoperability, and large municipalities are hindered by vendor lock-in and outdated technological frameworks. The study emphasizes the importance of context in e-health implementation, suggesting tailored strategies for municipalities based on size and capacity. This paper contributes to understanding the complexities of e-health implementation, offering insights for policymakers and practitioners aiming to optimize healthcare delivery through technology.**

***Keywords- E-Health Implementation; Municipal Healthcare; System Interoperability.***

## I. INTRODUCTION

E-health systems, defined as the use of Information and Communication Technology (ICT) to improve health services and outcomes, have become integral to modern healthcare delivery. Globally, these systems address pressing challenges such as aging populations, escalating healthcare costs, and the need for more efficient care delivery [1]. In Norway, where municipalities manage primary healthcare services, effective e-health implementation is vital for equitable and sustainable healthcare delivery [2].

The implementation of e-health systems in municipalities is influenced by a complex interplay of technical, organizational, and human factors. One prominent challenge is the fragmentation of healthcare information systems, which often lack interoperability and hinder seamless data exchange across different service levels and institutions [3]. This fragmentation is exacerbated by the varying capacities of municipalities to implement and manage advanced technological solutions, with smaller municipalities often struggling with limited financial and human resources [4]. This fragmentation is further complicated by a historical lack of regional ana national coordination in municipal e-health initiatives. Additionally, the dependence on proprietary systems has created issues such as vendor lock-in, which constrains innovation and limits municipalities' ability to tailor solutions to local needs [5].

Recognizing these challenges, various initiatives have in recent years emerged to support the adoption and integration of e-health systems. In Norway, the 2020 "Akson" project sought to establish a unified electronic health record (EHR) system for almost all municipalities, aiming to address interoperability issues and enhance data sharing (Directorate of eHealth, 2020). Similarly, a national e-health coordination network ("KS e-komp") directed by the municipalities' interest organization provides a collaborative platform for municipalities to share knowledge, resources, and best practices in e-health implementation. While these initiatives represent important steps forward, the effectiveness of such programs often depends on their ability to account for the diverse contexts and needs of municipalities of varying sizes.

From an academic perspective, research on e-health implementation has emphasized the importance of contextual factors, such as organizational readiness, governance structures, and user acceptance, in determining success [6][7]. However, there remains a gap in understanding how these factors differ across municipalities of different sizes and capacities. Addressing this gap is crucial for developing targeted policies and strategies that can enhance e-health adoption and performance across all levels of municipal healthcare.

In Section 2, we present the methodological approach of the study, including the research design, data collection, and analysis procedures. Section 3 outlines the results, structured around key themes identified in small, medium, and large municipalities. In Section 4, we discuss these findings in relation to existing literature, focusing on system challenges, collaboration, and attitudes toward change. Section 5 highlights the implications for policy and practice and concludes the paper by summarizing key insights and suggesting directions for future research.

## II. METHODS

This study adopts an interpretive research tradition to explore the size-dependent challenges and strategies associated with e-health implementation in Norwegian municipalities. The study employs a qualitative methodology, specifically using semi-structured interviews and thematic framework analysis, to gain an in-depth understanding of the phenomena under investigation.

The interpretive approach was chosen to investigate the nuanced and context-specific challenges faced by municipalities of varying sizes. This tradition allows for the exploration of participants' perspectives and experiences, emphasizing meaning-making processes in the context of e-health implementation. Interpretive research focuses on understanding phenomena within their specific contexts by engaging deeply with participants' lived experiences [8]. Such an approach is particularly well-suited for exploring the socio-technical dynamics of e-health, as it acknowledges the interplay between technology, organizational structures, and human behavior [9].

### *A. Data Collection*

Data were collected through semi-structured interviews conducted between April and June 2021. The interviews were designed to elicit detailed accounts of the challenges and strategies related to e-health systems in small, medium, and large municipalities.

### *B. Sampling strategy*

The study employed a purposive sampling strategy to recruit at least two informants from each municipality size category. The sample of six municipalities was chosen to capture variation across size categories, not to achieve representativeness. Purposive sampling ensured relevant informants, and two cases per category allowed for meaningful comparisons within the study's qualitative scope. While the sample is limited, it was sufficient to identify key themes and size-related differences. Classification followed the Statistics Norway (SSB) guidelines: municipalities with fewer than 5,000 inhabitants were categorized as "small," those with 5,000 to 19,999 as "medium," and those with over 20,000 as "large" [10]. A convenience list of four municipalities in the "small", and five in the "medium" categories, and two in the large category were created. Invitations to participate were sent to the "IT Director" or a similar role for municipalities that listed such contact information on their website. If no IT contact was listed, the invitation was directed to the "Director of Health Services" or equivalent. For municipalities without any specific contact details listed, invitations were sent to the official email address of the municipality. Of the eleven invitations sent, six municipalities accepted, one declined, and four did not respond. The six respondents were evenly distributed across the three municipality size categories.

### *C. Interview Process*

Interviews were conducted using video conferencing systems, ensuring flexibility and accessibility for participants across various municipalities. Each interview lasted between 45 and 90 minutes and was audio-recorded with the participants' consent. The recordings were subsequently transcribed verbatim to ensure an accurate representation of the data.

### *D. Data Analysis*

Thematic framework analysis was employed to identify patterns and themes within the data. This method provides a systematic approach to organizing and interpreting qualitative data while maintaining flexibility to accommodate emergent themes [11]. The analytical process consisted of several iterative steps to ensure depth and rigor in the interpretation of findings.

The analysis began with familiarization, during which the researchers reviewed the transcripts multiple times to immerse themselves in the data. Preliminary observations and reflections were recorded to capture initial impressions and identify potential areas of interest. Following this, coding was conducted inductively, with codes generated directly from the data rather than predefined categories. This inductive approach aligns with recommendations for thematic analysis, allowing the data to guide the development of key concepts [12]. For example, statements such as "Switching systems seems like a daunting process due to extensive re-training requirements" were coded as "Resistance to System Changes," while remarks like "We developed a platform to consolidate data from various systems during COVID" were coded as "Efforts Toward Unified Data Systems."

Once initial codes were generated, they were organized into thematic categories (Table 1), forming the basis of the analytical framework. This step involved grouping codes by their conceptual similarity and relevance to the research questions.

TABLE I. THEMATIC CATEGORIES

| *Category* | *Description* |
|---|---|
| E-health system challenges | Captures mentions of specific challenges within electronic health systems, including issues with integrations, user-friendliness, and system maintenance. |
| System implementation experiences | Relates to experiences and perspectives on implementing e-health systems, including the transition processes, user adoption, and training needs. |
| Stakeholder perspectives on system needs | Captures the expressed needs and requirements from various stakeholders like healthcare providers, system administrators, and local government officials regarding e-health systems. |
| Comparisons of system functionalities | Used for parts of the transcript where comparisons are made between different systems or where the functionality of existing systems is evaluated against potential new systems. |
| Innovations and improvements discussed | Any discussion related to innovations in e-health systems, potential improvements, or emerging technologies being considered for implementation. |
| Interactions with vendors and suppliers | Includes any mentions of interactions with technology vendors and suppliers, including negotiations, challenges, and the dynamics of vendor relationships. |
| Impact of e-health on clinical practices | References to how e-health systems impact clinical practices, decision-making, and patient management. |
| Policy and regulatory references | Any mention of policy, regulations, or national guidelines that influence the decision-making and operations of e-health systems. |
| Cross-municipal collaboration and challenges | Discussions related to the collaboration across different municipalities or regions, including shared challenges and collaborative projects. |

Table shows thematic categories developed from the initial codes

The category "E-health System Challenges" included codes such as "Integration Difficulties," "User-friendliness Issues," and "System Maintenance Challenges." In parallel, the category "System Implementation Experiences" encompassed codes like "Transition Processes," "User Adoption," and "Training Needs." This process of categorization was iterative, with the researchers continually revisiting and refining the groupings as new insights and patterns emerged from the data. As the analysis progressed, the framework was systematically applied across all transcripts to ensure consistency in identifying and interpreting the recurring themes. In total, we identified nine thematic categories that comprehensively captured the nuanced experiences and varying strategies employed by municipalities of different sizes in managing and innovating their e-health systems. These categories provided a robust framework for analyzing the diverse challenges and innovative solutions in e-health system implementation across municipal contexts.

The final step involved interpreting the themes in relation to the size of municipalities, emphasizing contextual differences and their implications for e-health implementation. This phase required synthesizing the themes to construct a coherent narrative that reflected the complexities of the data.

### E. Ethical Considerations

According to §2 and §4 of the Norwegian Act on Medical and Health Research, this study did not require approval from the Regional Ethics Committee (REK). The data handling procedures were approved by the Data Protection Officer at the University Hospital of North Norway. Participants provided informed consent prior to their involvement, including consent for audio recording and transcription. Confidentiality was ensured by anonymizing participant data and securely storing all research materials.

### F. Limitations

While the interpretive approach provides rich, contextual insights, the study's reliance on purposive sampling may limit the generalizability of its findings. For instance, larger municipalities typically have designated IT directors, whereas in smaller municipalities, these responsibilities often overlap with other roles and are not the primary focus. Consequently, there is a higher level of confidence in having identified the most suitable informants in larger municipalities. Additionally, the necessity of conducting interviews remotely due to pandemic restrictions during the study period may have impacted the depth of interaction with participants. The methodological approach allowed for an exploration of e-health implementation practices across municipalities, offering a basis for analyzing and understanding the challenges and strategies related to their sizes.

## III. Results

This section presents findings from the comparative analysis of e-health challenges and strategies across small, medium, and large municipalities in Norway. The results are structured around three primary themes: system challenges, collaboration and knowledge sharing, and attitudes toward change. Differences related to municipality size are highlighted, providing a nuanced understanding of the factors influencing e-health implementation.

### A. System Challenges

#### 1) Small Municipalities

Participants from small municipalities consistently highlighted the complexity and outdated nature of their e-health systems. These systems impose significant usability challenges on healthcare staff, many of whom lack the technical expertise required to navigate them effectively. Limited technical resources and staffing capacity further exacerbate these issues, necessitating reliance on external support for system management. For example, one respondent stated, "Gerica (their ehr system) is a bit outdated...a significant program that lacks intuitive design and requires substantial training." Another participant emphasized the burden of retraining, noting, "Switching systems seems like a daunting process due to extensive re-training requirements.".

#### 2) Medium Municipalities

In medium municipalities, integration and interoperability emerged as primary concerns. Participants reported difficulties in achieving seamless document sharing and aligning patient medication lists across systems. The persistence of fragmented systems was attributed to specialist groups favoring tailored software solutions, resulting in operational inefficiencies. The diversity of systems within these municipalities was noted as a complicating factor, hindering efforts to establish consistent workflows across healthcare services.

#### 3) Large Municipalities

Respondents from large municipalities identified vendor lock-in as a major challenge, emphasizing the constraints it imposes on data access and innovation. Large municipalities often depend on proprietary systems that limit their ability to integrate new functionalities or collaborate with external stakeholders. While smaller municipalities also faced challenges with proprietary applications, the more complex system requirements in large municipalities made this challenge more pressing in comparison. For instance, one respondent remarked, "We face significant vendor lock-in that prevents us from accessing our own data and innovating freely." Current platforms were described as technologically outdated and tightly coupled, making updates and modifications both slow and complex. Despite these challenges, respondents demonstrated an advanced awareness of data governance issues and expressed a strong desire to modernize system architectures.

### B. Collaboration and Knowledge Sharing

#### 1) Small Municipalities

Collaboration among small municipalities was primarily operational in nature, focusing on resource pooling to address capacity limitations. Shared servers, joint training initiatives, and regional cooperation were identified as essential strategies for managing e-health systems efficiently. Respondents highlighted the importance of these collaborative efforts in overcoming resource constraints and ensuring the continuity of system operations.

One participant shared, "We began with shared servers and system responsibilities among several municipalities." Another added, "Training and courses are shared among the municipalities to reduce resource strain.".

*2) Medium Municipalities*

Medium municipalities leveraged knowledge-sharing networks such as KS e-komp to align strategies and share best practices. These networks facilitated greater engagement with national initiatives, enabling municipalities to address system integration challenges more effectively. Collaboration in this context was more structured than in small municipalities, reflecting the increased complexity of e-health operations. One respondent described the benefits of such networks: "We work closely with other municipalities through KS e-komp to share knowledge and approaches." Another highlighted broader opportunities, noting, "Being part of national initiatives provides opportunities to align with broader goals."

*3) Large Municipalities*

Collaboration in large municipalities extended beyond operational and strategic alignment to include policy-level engagement. Respondents highlighted their involvement in shaping e-health policies and engaging with national and international stakeholders. The COVID-19 pandemic served as a catalyst for innovation, with several large municipalities developing platforms to consolidate data from fragmented systems. These initiatives demonstrated the potential of collaboration to address systemic challenges effectively.

One participant explained, "We developed a platform to consolidate data from various systems during COVID, which has been well-received."

### C. Attitudes Toward Change

*1) Small Municipalities*

Respondents from small municipalities expressed significant resistance to change, driven by concerns over resource limitations and the perceived burden of system adaptation. Fear of disruptions to established workflows and the high cost of training further contributed to this resistance. Change was often viewed as a risk rather than an opportunity. For example, one participant noted, "Switching systems seems like a daunting process due to extensive re-training requirements." Another stated, "Basic knowledge is not enough to handle the system effectively; it feels like one needs an IT background."

*2) Medium Municipalities*

In medium municipalities, change was generally perceived as necessary but was hindered by fragmentation and conflicting preferences among specialist groups and departments. The lack of a unified approach to system adoption posed additional challenges, complicating efforts to implement large-scale changes effectively. One respondent commented, "Fragmentation persists as specialist groups favor their specific software solutions." Another highlighted, "The lack of integration makes large-scale changes very challenging."

*3) Large Municipalities*

Respondents from large municipalities exhibited a proactive attitude toward transformative change, driven by the need to address data governance and improve interoperability. However, external factors such as vendor dependencies and systemic bottlenecks at the national level constrained their ability to implement these changes. Despite these limitations, large municipalities demonstrated a strong commitment to innovation and modernization. One respondent emphasized, "We are constrained by vendor dependencies, but we remain committed to driving innovation and modernization." Another explained, "Transformative changes require national-level coordination, which can be slow."

## IV. Discussion

The findings from this study highlight the size-dependent challenges faced by municipalities in implementing e-health systems. By examining these challenges and the corresponding strategies, this discussion contextualizes the contingencies of municipality size in shaping e-health capabilities and outcomes.

### A. System Challenges Across Municipalities

The analysis underscores how municipality size shapes the scope and nature of e-health challenges. Small municipalities face fundamental operational difficulties, such as outdated systems and limited technical capacity. These findings align with prior research suggesting that rural and smaller local governments often lack the infrastructure and human resources needed for technological innovation [4]. This reliance on external support and resistance to system changes, driven by resource constraints, underscores the importance of designing low-complexity, user-friendly e-health systems. For example, as one respondent stated, "Switching systems seems like a daunting process due to extensive re-training requirements." Such insights align with findings by Venkatesh et al. [6], who emphasize the critical role of perceived ease of use in technology adoption.

In contrast, medium municipalities encounter challenges primarily related to integration and interoperability. The coexistence of multiple, fragmented systems tailored to specific functions hinders seamless data sharing and workflow consistency. These findings resonate with studies highlighting the complexities of integrating heterogeneous systems in mid-sized organizations [13]. Addressing this issue requires technical solutions, such as standardized data exchange protocols, and governance mechanisms to align diverse stakeholders [14].

Large municipalities contend with advanced challenges, such as vendor lock-in, which restricts access to data and limits innovation. These municipalities operate on outdated platforms that are tightly coupled, making changes both slow and resource-intensive. For instance, that significant vendor lock-in prevented access to their own data, challenging innovation. The persistence of vendor lock-in reflects broader issues in e-health system procurement, where proprietary solutions often dominate. These findings suggest that large municipalities would benefit from national-level interventions to promote open data standards and reduce dependency on proprietary systems [15]. Municipalities may benefit from exploring open standards and open-source solutions like openEHR, which can support data portability and reduce dependency on single vendors. Joint procurement initiatives could also offer greater leverage in negotiations. Additionally, national policies that encourage interoperability and vendor-neutral approaches might help create more flexible and sustainable systems over time.

### *B. Collaboration and Knowledge Sharing*

Collaboration emerges as a critical strategy across municipalities, but its nature and purpose vary with size. Small municipalities rely on collaboration to pool resources and share operational responsibilities. Shared training and infrastructure initiatives allow them to overcome capacity constraints. This observation aligns with studies highlighting the efficacy of intermunicipal cooperation in addressing resource disparities [16].

Medium municipalities utilize collaborative networks such as KS e-komp to align strategies and share best practices. These networks enable municipalities to collectively address integration challenges and benefit from national-level guidance. "We work closely with other municipalities through KS e-komp to share knowledge and approaches," explained one respondent. The role of such networks emphasizes the importance of collaborative innovation in public sector reforms.

In large municipalities, collaboration extends beyond operational needs to policy-driven initiatives and innovation. Their involvement in shaping e-health policies at national and international levels underscores their strategic importance. For instance, during the COVID-19 pandemic, several large municipalities developed platforms to consolidate fragmented data systems. This innovation exemplifies the transformative potential of policy-aligned collaboration.

### *C. Perspectives on Change*

The willingness and ability to innovate are influenced by municipality size. Small municipalities exhibit cautious attitudes toward change, driven by resource limitations and concerns over workflow disruptions. This risk aversion aligns with studies suggesting that smaller organizations are slower to adopt new technologies due to financial and technical constraints [4].

Medium municipalities display moderate openness to change but encounter internal challenges related to fragmentation and stakeholder misalignment. Effective governance structures are crucial for fostering coordinated efforts toward e-health improvements.

Large municipalities adopt a proactive approach to innovation, motivated by the need to address advanced challenges like data governance and interoperability. However, progress is constrained by external factors, including vendor dependencies and national policy bottlenecks. Aligning local efforts with broader systemic reforms is essential to overcoming these barriers [17].

### *D. Implications for Policy and Practice*

The insights from this study underscore the importance of context-sensitive strategies in e-health implementation. Policies must be tailored to the unique challenges and capacities of municipalities of varying sizes. For small municipalities, low-complexity and cost-effective systems are essential to enhance usability and minimize resource strain. Medium municipalities require interoperable solutions and governance mechanisms to address system fragmentation. Large municipalities need systemic support to overcome vendor lock-in and implement open data standards, enabling innovation and modernization.

Strengthening regional and national collaboration frameworks can further support municipalities in addressing shared challenges. Initiatives that align strategies and facilitate knowledge sharing are instrumental in fostering innovation and improving e-health outcomes. Capacity-building programs targeted at smaller municipalities can mitigate resistance to change and enhance digital competency, as evidenced by successful training initiatives in public sector organizations [18].

## V. CONCLUSION

This study highlights the size-dependent challenges and strategies associated with e-health implementation in Norwegian municipalities. Small municipalities struggle with limited resources and outdated systems, medium municipalities face issues of integration and interoperability, and large municipalities grapple with vendor lock-in and data governance challenges. By examining these differences, this study underscores the importance of tailoring e-health policies and initiatives to the specific needs of municipalities based on their size and capacity. These findings provide valuable insights for policymakers and practitioners, emphasizing the need for inclusive and context-sensitive strategies to ensure equitable and effective e-health adoption.

## REFERENCES

[1] World Health Organization, "Global strategy and action plan on ageing and health," https://iris.who.int/bitstream/handle/10665/329960/9789241513500-eng.pdf 2017.

[2] Å. Ringard, A. Sagan, I. S. Saunes, and A. K. Lindahl, "Health systems in transition," *Health*, vol. 15, no. 8, pp. 1-162, 2013.

[3] K. Malm-Nicolaisen, R. Pedersen, and A. J. Fagerlund, "Exploring the Emergence of Open Platforms in Healthcare: Design Considerations and Experiences from an Initial Case in Norwegian Primary Care," *(Pre-press)*, 2022.

[4] E. Hage, J. P. Roo, M. A. van Offenbeek, and A. Boonstra, "Implementation factors and their effect on e-Health service adoption in rural communities: a systematic literature review," *BMC health services research*, vol. 13, pp. 1-16, 2013.

[5] K. Malm-Nicolaisen, R. Pedersen, and A. J. Fagerlund, "Measuring Perceptions of Openness in Health Information Technology Platforms: Results from Pilot Testing Proposed Survey Framework," in *eTELEMED*, pp. 26-32, 2022.

[6] V. Venkatesh, M. G. Morris, G. B. Davis, and F. D. Davis, "User acceptance of information technology: Toward a unified view," *MIS quarterly*, pp. 425-478, 2003.

[7] T. Greenhalgh *et al.*, "Beyond adoption: a new framework for theorizing and evaluating nonadoption, abandonment, and challenges to the scale-up, spread, and sustainability of health and care technologies," *Journal of medical Internet research,* vol. 19, no. 11, p. e8775, 2017.

[8] G. Walsham, "Interpretive case studies in IS research: nature and method," *European Journal of information systems,* vol. 4, no. 2, pp. 74-81, 1995.

[9] W. J. Orlikowski and J. J. Baroudi, "Studying information technology in organizations: Research approaches and assumptions," *Information systems research,* vol. 2, no. 1, pp. 1-28, 1991.

[10] M. Kringlebotten and A. Langørgen, *Gruppering av kommuner etter folkemengde og økonomiske rammebetingelser 2020.* Statistisk sentralbyrå, 2020.

[11] N. K. Gale, G. Heath, E. Cameron, S. Rashid, and S. Redwood, "Using the framework method for the analysis of qualitative data in multi-disciplinary health research," *BMC medical research methodology,* vol. 13, no. 1, pp. 1-8, 2013.

[12] V. Braun and V. Clarke, "Using thematic analysis in psychology," *Qualitative research in psychology,* vol. 3, no. 2, pp. 77-101, 2006.

[13] G. Ellingsen, M. Hertzum, and L. Melby, "The Tension between National and Local Concerns in Preparing for Large-Scale Generic Systems in Healthcare," *Computer Supported Cooperative Work (CSCW),* pp. 1-31, 2022.

[14] K. Malm-Nicolaisen, L. M. Ruiz, E. R. Evenstad, and R. Pedersen, "Efforts on Using Standards for Defining the Structuring of Electronic Health Record Data: A Scoping Review," in *SHI 2019. Proceedings of the 17th Scandinavian Conference on Health Informatics, November 12-13, 2019, Oslo, Norway*, 2019, no. 161: Linköping University Electronic Press, pp. 108-115.

[15] G.-H. Ulriksen, R. Pedersen, and G. Ellingsen, "Infrastructuring in healthcare through the openEHR architecture," *Computer Supported Cooperative Work (CSCW),* vol. 26, no. 1-2, pp. 33-69, 2017.

[16] R. Hulst and A. v. Montfort, "Inter-municipal cooperation: A widespread phenomenon," in *Inter-municipal cooperation in Europe*: Springer, 2007, pp. 1-21.

[17] M. B. Lips, "Transforming government services over time: Meanings, impacts, and implications for citizen-government relationships," in *Routledge Handbook on Information Technology in Government*: Routledge, 2017, pp. 11-26.

[18] L. K. Roland, T. A. Sanner, and M. Aanestad, "Flexibility in EHR ecosystems: five integration strategies and their trade-offs," in *Norsk konferanse for organisasjoners bruk at IT*, 2017, vol. 25, no. 1.

# Evaluating Usability Barriers in Health Technology: Government Perspectives and ISO Standard Alignment

Marianne Lodvir Hemsing
Department of Business, Strategy and Political Science
University of South-Eastern Norway
Kongsberg, Norway
e-mail: mlhemsing@gmail.com

*Abstract*— **This study investigates how government authorities in Norway perceive usability barriers in health technology and assesses their alignment with international usability and software quality standards. A content analysis of 15 government reports reveals that the most frequently cited barriers—system complexity, accessibility issues, and integration challenges—closely align with key usability requirements in these standards. The findings suggest that adopting these usability principles as guidelines in health technology development could enhance effectiveness and accessibility. This study underscores the importance of structured usability frameworks in addressing real-world challenges.**

***Keywords-Usability in Health Technology; Standards; ISO 9241-11; ISO/IEC 25010; Government Policy in Digital Health.***

## I. INTRODUCTION

Existing research has provided valuable insights into how healthcare teams interact with digital tools, emphasizing the need for seamless integration and user-friendly design to support clinical workflows and patient care [1]. However, the usability of health technology systems remains a major challenge [2]. There are continued usability challenges in health technologies, affecting both patients and healthcare professionals. Research shows that usability flaws can lead to medical errors, patient harm, and frustration among clinicians, contributing to burnout [2][3].

Government plays a crucial role in the success of health technology [4]. Government agencies are often responsible for funding early-stage innovations, setting regulatory standards, and coordinating between public and private sectors to ensure the smooth deployment of digital health solutions. Unlike research papers, which aim to contribute new knowledge to an academic field, government reports prioritize practical information and real-world applications. By analyzing the government's perceived usability barriers, we aim to look at the problem from a top-down perspective and get a better understanding of whether the industry standards meet the government's perceived need.

The two standards from the International Standardization Organization (ISO) and International Electrotechnical Commission (IEC), evaluated in this paper, are widely used in the fields of user experience design, software development, and systems engineering to ensure high-quality, user-friendly, and effective systems:

*1) ISO 9241-11:2018: Ergonomics of human-system interaction – Part 11: Usability: Definitions and concepts* [5]. This standard focuses on usability, providing definitions, principles, and concepts for understanding how to assess the usability of systems, including software and hardware. It emphasizes the importance of usability in terms of user effectiveness, efficiency, and satisfaction when interacting with a product or system.

*2) ISO/IEC 25010:2023: Systems and software engineering — Systems and software Quality Requirements and Evaluation (SQuaRE) — Product quality model* [6]. This standard provides a framework for evaluating the quality of software and systems. The models include various characteristics like functionality, reliability, usability, and maintainability, which are used to assess and ensure software quality. Usability is framed within the concept of interaction capability (a shift from earlier versions that treated usability as an independent characteristic). The standard defines usability in terms of how effectively, efficiently, and satisfactorily a system or product can help users achieve their goals in a specified context. It includes six sub-characteristics that detail usability requirements:

- Appropriateness
- Recognizability: Users can quickly identify if the system meets their needs.
- Learnability: The system should be easy for users to learn, reducing the time and effort required to start using it effectively.
- Operability: Attributes that make the system easy to control and operate.
- User Error Protection: Mechanisms that help prevent user mistakes or mitigate their impact.
- User Assistance: Systems should offer guidance and support features to aid users in understanding and using them effectively.
- Accessibility: Ensuring usability across a diverse range of users, including those with disabilities.

Previous research has explored how the ISO 9241-11 standard is applied in real-world settings in healthcare. In [7], the authors highlight both the benefits and challenges of using ISO 9241-11 for setting usability targets. The findings in the article show that while the standard provides a solid foundation for usability, determining the right levels of effectiveness, efficiency, and satisfaction can be complex, and requires further methodological support. Another relevant research paper explores how the ISO/IEC 25010:2011 standard is applied to medical device software, and highlights the importance of qualities such as usability, reliability, security, and maintainability [8].

The remainder of this paper is structured as follows: In Section 2, the methodology is described, in Section 3, the results are presented together with recommendations for future research, and in Section 4 we draw the conclusions.

## II. Methodology

This research employs a hybrid content analysis approach to systematically examine the usability and quality requirements discussed in reports published by Norwegian government authorities. 15 reports were selected for analysis, all sourced from public authorities in Norway, and published between 2021 and 2024. The documents include government reports, healthcare surveys, technology assessments, and strategy papers on digital health and Artificial Intelligence (AI) in clinical settings. The reports have a total of 366,611 words, equivalent to between 1200 and 1500 pages.

The process of content analysis was based on extracting meaningful information from the reports using predefined themes and keywords, which were based on the concepts laid out in the two ISO standards. Themes and keywords were selected through iterative reading of the reports and validated against the terminology used in the standards to ensure alignment. Both manual and automated techniques were used to ensure comprehensive coverage of the reports.

In addition, a manual in-depth analysis of the content in the reports and the requirements in the two ISO standards were conducted.

### A. *Quantitative and Qualitative Content Analysis*

To systematically analyze the documents, a set of keywords and themes were defined in advance. These keywords were derived from the critical components of ISO 9241-11:2018 and ISO/IEC 25010:2023 and were supplemented by terms identified as relevant to health technology usability in the reports. The keywords were grouped under the following themes: Security, Efficiency, Satisfaction, Learnability, Usability, Accessibility, User Experience, User Engagement, Interoperability, Technical Barriers.

The analysis involved both manual review and the use of automated analysis to extract text, count keyword occurrences, and identify recurring themes across the documents. A Chi Square Test was used to evaluate distribution of the themes. This provided a frequency-based view of the data, allowing for insights into which topics were most emphasized in relation to usability and system quality. These frequencies were also compared across the reports to identify patterns, such as which aspects of usability or quality were prioritized in different contexts.

After identifying the key themes and barriers, a manual in-depth analysis of the content in the reports and the requirements in the two ISO standards where conducted, the findings from the reports were systematically mapped to the relevant principles from both ISO standards. The dataset is available online [9].

## III. Results

A main finding from the analysis is that all the reports described usability or related topics such as ease of use or user-centered design in the context of health technologies, though the frequency and depth of these discussions vary. Data Saturation was achieved even though the sample was limited. This shows that the government consistently recognizes usability as an important factor when discussing health technology.

The Chi Square Test in Table I with a p-value of 0 demonstrates that the themes are not equally distributed. Certain themes (like security) are disproportionately emphasized, while others (like technical barriers) are underrepresented. This supports the idea that specific aspects of usability (e.g., security, efficiency) receive far more attention than others, which could influence the focus of health technology development. Security is not analyzed further as the focus is usability, and security is a substantial area that would require separate analysis and evaluation against other standards, such as the ISO/IEC 27000 family of standards on information security management.

TABLE I. Chi Square Results

| Theme Code | Observed | % | Expected* | Chi-Square | p-Value |
|---|---|---|---|---|---|
| Security | 581 | 36 | 162.4 | **1980.67** | 0 |
| Efficiency | 383 | 24 | | | |
| Satisfaction | 204 | 13 | | | |
| Learnability | 203 | 12 | | | |
| Usability | 112 | 7 | | | |
| Accessibility | 60 | 4 | | | |
| User Experience | 37 | 2 | | | |
| User Engagement | 23 | 1 | | | |
| Interoperability | 16 | 1 | | | |
| Technical Barriers | 5 | 0 | | | |

Degrees of freedom: 9, *value for equal distribution

### A. *Reccuring Themes*

The main recurring themes were identified as follows:

*1) User-Centered Design:*

Nearly all reports stress the importance of designing health technologies with the end-users in mind, whether they are healthcare professionals or patients. Usability in this context means creating interfaces and workflows that are intuitive, reducing cognitive load, and ensuring that users can achieve their objectives effectively. This is strongly tied to the ISO 9241-11 emphasis on the effectiveness, efficiency, and satisfaction of users.

*2) Seamless Integration and Interoperability:*
Usability challenges are frequently linked to issues of system interoperability, where healthcare professionals struggle to interact efficiently with multiple systems that do not communicate well. This issue is critical for effective digital health solutions, and usability suffers when systems require users to navigate disjointed interfaces.

*3) Ease of Use for Healthcare Providers:*
Usability for healthcare providers, specifically in terms of reducing complexity in accessing, documenting, and sharing patient information, is a recurring theme. Many reports call for technologies that are simple, easy to learn, and support fast decision-making, directly corresponding to ISO/IEC 25010:2023's emphasis on usability attributes such as learnability, operability, and user error control.

*4) Patient Empowerment and Accessibility:*
Many reports emphasize the need for health technologies to be usable not just by healthcare professionals but by patients as well. This includes creating interfaces that are accessible, ensuring inclusivity for users with different abilities and technical proficiencies. Making systems accessible to a broad user base also aligns with ISO standards, which emphasize user satisfaction and accessibility.

*5) Error Reduction and Safety:*
Usability is closely linked to patient safety, as emphasized in the reports. Poor usability can lead to errors, particularly in clinical settings where systems that are difficult to navigate or understand can cause mistakes in treatment or data entry. ISO standards also emphasize the error control aspect of usability, ensuring that systems minimize the potential for user errors.

*6) Training and Support for Users:*
Several reports highlight the need for comprehensive training and ongoing support for users to maximize the usability of health technology systems. This is particularly important for systems with steep learning curves, where even well-designed interfaces can be difficult to use without proper instruction.

*7) Adaptability and Customization:*
Health technologies are increasingly required to offer customizable options for different user groups (e.g., different levels of healthcare professionals, patients). Usability is enhanced when users can adapt systems to their specific workflows or preferences, making them more efficient and effective in their tasks.

In summary, user-centered design, system integration, ease of use for healthcare professionals, patient accessibility, error reduction, training and support, and adaptability are recurring usability themes across the reports, and are thus perceived as the most important components in a government perspective. These themes highlight the essential qualities that health technologies need to embody to align with ISO standards for usability.

### *B. Government Perceived Barriers*

Challenges and barriers related to usability in health technologies were identified as follows:

*1) Digital Divide and Accessibility Issues:*
Access to technology and digital literacy are highlighted as significant barriers, particularly for vulnerable populations such as the elderly, low-income groups, or those with limited technical skills. These populations struggle with accessing and effectively using digital health services. In reports discussing patient portals, many individuals from disadvantaged backgrounds report difficulties using these systems due to insufficient training or lack of access to necessary devices.

*2) User Interface Complexity:*
Many users, both patients and healthcare providers, report that complex or unintuitive interfaces in health technologies like patient portals, EPJ systems, and telemedicine platforms hinder effective use. Systems with overly complex designs reduce the ease of use, especially in high-pressure healthcare environments. Healthcare providers have highlighted challenges in using systems that require extensive training to navigate, reducing efficiency and increasing errors.

*3) Usability for Vulnerable Populations:*
Vulnerable groups, including those with disabilities or chronic illnesses, report that many health technologies are not designed with their specific needs in mind, leading to usability issues. For example, individuals with chronic diseases or disabilities find it difficult to use mobile health apps that do not accommodate specific needs. Lack of accessibility features (e.g., for visually impaired users) and poor adaptability to different patient needs are recurrent themes.

*4) Interoperability and Integration Challenges:*
A lack of interoperability between different systems (e.g., between hospitals, clinics, and patient devices) makes it harder for healthcare providers to access relevant information, which in turn affects user experience and the overall usability of health technology. Reports frequently mention difficulties in integrating telemedicine platforms with other health systems, causing fragmented workflows and data management issues.

*5) Technical Barriers and System Downtime:*
System downtime and technical issues are frequently mentioned as barriers to the usability of health technology. These interruptions not only reduce user trust but also compromise patient care and data access. Users express frustration when health systems experience downtime or fail to deliver expected results efficiently during clinical operations.

*6) User Trust and Satisfaction:*
Lack of user trust in digital solutions is another barrier to widespread adoption. Users, both healthcare providers and patients, often feel unsure about the reliability of new technologies, which limits engagement and satisfaction. Patients often feel disconnected or uncertain when using telehealth services due to inadequate technical support or system reliability.

These insights show that while digital health technologies offer significant potential, there are numerous usability barriers that need to be addressed for better adoption and effectiveness. Solutions, such as improving system interfaces, enhancing interoperability, and ensuring systems are designed with vulnerable populations in mind will help address these challenges. Table II summarizes how the two ISO standards can aid in solving the recurring barriers.

TABLE II. RECURRING BARRIERS

| Barrier | ISO 9241-11:2018 | ISO/IEC 25010:2023 | Can ISO Standards Solve This? |
|---|---|---|---|
| **Digital Divide and Accessibility Issues** | Emphasizes context of use, ensuring that systems are designed for a diverse range of users, including those with limited digital skills or access to technology. | ISO/IEC 25010:2023 includes accessibility as a sub-characteristic of usability, pushing for systems that cater to the needs of vulnerable populations, such as the elderly and disabled. | Partially. The standards provide principles for accessible design, but the digital divide often stems from external factors, such as lack of access to devices or internet connectivity, which standards alone cannot address. Government policies or broader infrastructure improvements are needed alongside these usability guidelines. |
| **User Interface Complexity** | Focuses on effectiveness and efficiency, ensuring that systems are designed to be intuitive and easy to navigate | Includes operability and learnability as critical components of usability, ensuring that systems should be easy to learn and operate, minimizing the cognitive load for users. | Yes, largely. If applied rigorously during the design and evaluation stages, these standards can significantly reduce interface complexity by enforcing user-centered design and ensuring that systems are intuitive and straightforward to use. |
| **Usability for Vulnerable Populations** | Specifically considers the context of use for different user groups, which can ensure that systems are designed to be inclusive of vulnerable populations. | Emphasizes accessibility and user satisfaction, pushing for systems that are not only functional but also usable for people with disabilities or chronic conditions. | Yes, to a significant extent. If developers follow these guidelines, they can create inclusive designs that accommodate the needs of vulnerable groups. However, this requires a commitment from developers to prioritize accessibility and ensure that systems are tested by diverse user groups during the design phase. |
| **Interoperability and Integration Challenges** | | Includes compatibility and interoperability as critical quality characteristics, ensuring that systems can work together without causing usability issues. | Partially. While the standards promote interoperability, they cannot fully solve integration challenges caused by legacy systems, incompatible infrastructure, or organizational issues. To achieve seamless integration, there must be broader cooperation between vendors, developers, and healthcare institutions to implement systems that follow standardized data formats and communication protocols. |
| **Technical Barriers and System Downtime** | | Addresses reliability by requiring systems to minimize downtime and ensure that recovery from failures is efficient. | Partially. The standards can push for more reliable systems, but issues like system downtime are often due to infrastructure, network failures, or inadequate resources. While following standards can help reduce technical problems, solving them completely often requires investments in IT infrastructure and system maintenance. |
| **User Trust and Satisfaction** | Includes user satisfaction as a core component of usability, ensuring that systems are designed to meet user expectations and needs. | Emphasizes user satisfaction and trust through consistent reliability, security, and usability characteristics. | Yes, to a large extent. By focusing on usability, security, and reliability, ISO standards can improve user trust in health technologies. However, achieving trust also requires good communication, training, and support, which go beyond what the standards themselves prescribe. |

| Barrier | ISO 9241-11:2018 | ISO/IEC 25010:2023 | Can ISO Standards Solve This? |
|---|---|---|---|
| **Training and Support Needs** | Stresses the importance of learnability, meaning that systems should be easy to learn and use. This aligns with the need for less training if systems are inherently user-friendly. | | While standards can help design learnable systems, the need for ongoing training and support depends on how well healthcare institutions implement the technologies and provide resources for users. Standards cannot replace the need for user education, but they can reduce the complexity that necessitates heavy training. |

*C. Seven key usability improvements*

Based on the results from the content analysis, we suggest highlighting seven reoccurring key usability improvements shown in Table III. The table summarizes how these highlighted key improvements map to the requirement in the two ISO Standards:

TABLE III. KEY IMPROVEMENTS TO USABILITY

| Usability Improvement Suggestions | ISO 9241-11 (Usability) | ISO/IEC 25010 (Quality Model) |
|---|---|---|
| **1. Improve User-Centered Design** | Aligns with Effectiveness, Satisfaction, and Context of Use. User-centered design ensures that the system meets user needs and is effective in various contexts. | Maps to Usability (Subcharacteristics: Learnability, Operability, User error protection, User interface aesthetics). Focus on making systems easy to use and visually intuitive. |
| **2. Enhanced Training and Support** | Relates to Satisfaction and Efficiency. Users who are well-trained can achieve tasks more efficiently, leading to higher satisfaction. | Falls under Usability (Learnability) and Maintainability (Subcharacteristic: Modifiability). Well-supported systems improve long-term usability and adaptability. |
| **3. Simplify Interfaces and Reduce Complexity** | Focuses on Effectiveness and Efficiency. Simplified interfaces help users complete tasks more quickly and accurately, leading to improved usability. | Related to Usability (Operability) and Functional Suitability (Subcharacteristic: Functional appropriateness). The system should provide necessary functions in a way that users can easily access and understand. |
| **4. Increase System Interoperability** | Related to Context of Use, ensuring that systems work smoothly in various environments and contexts without creating barriers for the user. | Primarily aligns with Compatibility (Subcharacteristics: Interoperability, Co-existence). Systems need to work together seamlessly to prevent usability issues caused by fragmented workflows. |
| **5. Accessibility for Vulnerable Populations** | Related to Satisfaction and Context of Use. Ensuring usability for a diverse range of users, including those with disabilities or limited digital literacy. | Falls under Usability (Accessibility). The system must be accessible to users with a range of abilities and needs, ensuring inclusivity and equal access. |
| **6. Strengthen Feedback Loops** | Involves Measurement and Evaluation of Usability, ensuring that user feedback is continuously collected and used to improve usability. | Related to Maintainability (Subcharacteristic: Modifiability). Continuous feedback allows the system to be adjusted and improved over time. It also ties into Usability (User error protection and User satisfaction). |
| **7. Provide Clearer Guidelines for Digital Health** | Falls under Effectiveness, ensuring that new systems are assessed for their usability in meeting user goals. | Maps to Functional Suitability and Usability. Guidelines ensure that systems meet functional needs and provide the right level of usability, ensuring Reliability and Quality from the outset. |

### D. *Limitions and Future Research*

This study is based on 15 Norwegian government reports, which may limit the generalizability of findings to other health systems or user contexts. The focus on policy-level documents also excludes perspectives from patients, clinicians, or developers. Additionally, the use of ISO 9241-11 and ISO/IEC 25010, while widely accepted, may overlook alternative frameworks relevant to specific healthcare settings.

Future research could include broader stakeholder input, cross-country comparisons, or real-world testing of ISO standards in health technology design and evaluation. Future research should also explore how these standards can be operationalized in national strategies, and how policymakers can support more consistent usability evaluation across public health projects.

## IV. Conclusion

The findings in this content analysis show that the government perception on barriers and improvements in relation to usability in health technologies, align closely with the usability requirements set in ISO 9241-11:2018 "Ergonomics of human-system interaction – Part 11: Usability" and ISO/IEC 25010:2023 " Systems and software engineering — Systems and software Quality Requirements and Evaluation (SQuaRE) — Product quality model".

From this, we draw two conclusions:

- The requirements in ISO 9241-11:2018 and ISO/IEC 25010:2023 are highly relevant to the practical challenges faced in Norway.
- It would make a great deal of sense to use ISO 9241-11 and ISO/IEC 25010:2023 as requirements or guidelines for developing and accessing health technologies in Norway.

The findings suggest that the standards can serve as practical tools for shaping digital health policies by providing structured usability requirements that align with real-world challenges identified in national reports. Integrating these standards into policy frameworks and procurement processes could improve the quality, accessibility, and adoption of health technologies

However, the standards cannot fully solve external challenges like the digital divide, infrastructure issues, or organizational hurdles related to system integration. These standards, when properly implemented alongside broader efforts, can significantly improve the usability and quality of health technologies.

## References


[1] M. Sujan, "Integrating digital health technologies into complex clinical systems," BMJ Health & Care Informatics, vol. 30, no. 1, p.e100885, Oct 2023.

[2] R. Ratwani, J. Reider, and H. Singh, "A Decade of Health Information Technology Usability Challenges and the Path Forward" JAMA, vol. 321, no 8, pp. 743–744, Feb 2019.

[3] P. Carayon and P. Hoonakker, "Human Factors and Usability for Health Information Technology: Old and New Challenges," Yearbook of Medical Informatics, vol, 28, no 1, pp. 071–077, Aug 2019.

[4] K. Benjamin and H.W. Potts, "Digital transformation in government: Lessons for digital health?" DIGITAL HEALTH, vol 4, pp. 1-5, Feb 2018.

[5] ISO, "ISO 9241-11:2018, Ergonomics of human-system interaction – Part 11: Usability: Definitions and concepts".

[6] ISO, "ISO/IEC 25010:2023 Systems and software engineering — Systems and software Quality Requirements and Evaluation (SQuaRE) — Product quality model".

[7] M. Teruel, E. Navarro, V. López-Jaquero, F. Montero, and P. González, "A CSCW Requirements Engineering CASE Tool: Development and usability evaluation," Information and Software Technology, vol 56, no 8, pp. 922-949, Aug 2014.

[8] S. Han, R. Sinha, and A. Lowe, "Assessing Support for Industry Standards in Reference Medical Software Architectures," Proceedings of the 46th Annual Conference of the IEEE Industrial Electronics Society (IECON2020), pp. 3403-3407, Jul 2020.

[9] Figshare Dataset, 10.6084/m9.figshare.28335071.

# Method for Retrospective Analysis of Clinical Outcomes After Removal of Radiologist Assessment of AI-Positive Skeletal X-rays

Kristian Malm-Nicolaisen
Norwegian Centre for E-health Research
Tromsø, Norway
Email: kristian.nicolaisen@ehealthresearch.no

Rune Pedersen
Norwegian Centre for E-health Research
Tromsø, Norway
Email: rune.pedersen@ehealthresearch.no

Asbjørn Johansen Fagerlund
Norwegian Centre for E-health Research
Tromsø, Norway
Email: asbjorn.johansen.fagerlund@ehealthresearch.no

*Abstract*— **This retrospective in-progress study presents a method for evaluating the clinical impact of removing routine radiologist assessments of AI-flagged ("AI-positive") skeletal X-rays at Vestre Viken Health Trust, which recently implemented the BoneView AI system. Over a five-month period, orthopedic surgeons or radiologists prospectively identified False Positives (FPs), and comprehensive chart reviews assessed whether omitting radiologist input compromised patient safety or led to unnecessary interventions. In the current phase, an estimated 20-40 FP cases will be analyzed for treatment outcomes, resource utilization, and potential misdiagnoses. The findings will inform evidence-based strategies for integrating AI into radiological workflows, guiding institutions in balancing efficiency with the need for robust diagnostic oversight.**

***Keywords-artificial intelligence; diagnostic imaging; clinical workflow; patient outcomes.***

## I. INTRODUCTION

Artificial Intelligence (AI)-driven technologies hold significant promise in addressing critical challenges within healthcare, including workforce shortages and the increasing demands of an aging population with complex medical needs. Despite the proliferation of commercial AI solutions globally, the integration and deployment of these technologies in clinical environments remain limited [1]. While commercial AI algorithms undergo validation for clinical performance prior to market entry, early adopter institutions must determine how to integrate these solutions into clinical workflows and evaluate the implications of workflow modifications.

A notable example is Vestre Viken Health Trust in Norway, part of the public healthcare sector, which implemented BoneView (by Gleamer) across all its hospitals in 2023. BoneView is an AI-powered tool designed to assist radiologists in the detection and assessment of fractures in X-ray examinations. By embedding BoneView into routine radiological practice, Vestre Viken aims to enhance diagnostic accuracy, reduce radiologist workload, and improve patient care through faster and more reliable fracture assessments [2]. Throughout the implementation phase, Vestre Viken has continued its legacy workflow where radiologists assess clinical images that are labeled as diagnostically positive by the AI. The so-called "AI-positive" patients are triaged to orthopedics for clinical procedures and treatment. The radiology assessment thus often occurs *after* the orthopedic procedure is complete (see Figure 1). The added clinical value of the post-procedure radiology assessment is unknown, and the hospital wants to assess whether it is viable to end the practice of radiologist/AI double-assessment of *non-complex* AI-positive cases, to free radiology resources for more complicated tasks.

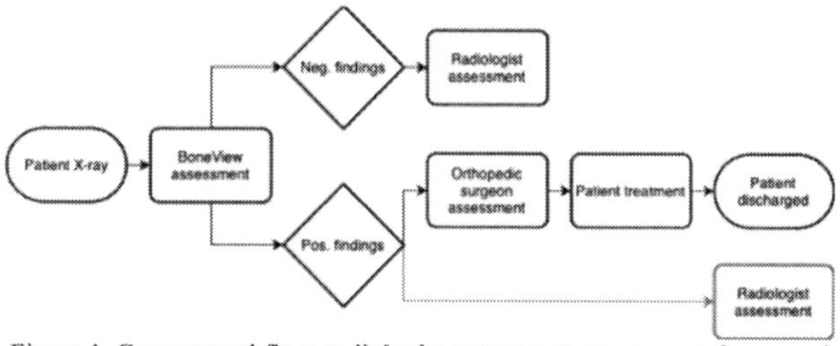


Figure 1. Current workflow, radiologist assessment step proposed removed highlighted in red.

To evaluate the clinical impact of this workflow modification in a live clinical environment, this paper outlines a retrospective observational study in progress. The study aims to determine the added clinical value of radiologists' assessments for AI-positive images by focusing on False Positive (FP) AI results identified through orthopedic or radiologic evaluation. Expert reviews of clinical documentation will further clarify the consequences for patient care, offering insights into whether radiologist oversight remains essential in AI-driven fracture detection.

The paper is structured as follows: Section 2 discusses AI integration into radiology workflows and the implications of removing radiologist assessments. Finally, Section 3 describes the study methodology, including data collection, sampling, analyses, and ethical considerations. Section 4 outlines the expected outcomes and their impact on clinical practice.

## II. Background

AI integration in radiology has traditionally followed a "radiologist-in-the-loop" model, where AI serves as a supportive tool rather than an autonomous decision-maker [3]. However, recent discussions have explored whether low-complexity cases (such as straightforward fractures) could be managed without radiologist reassessment to optimize resource utilization. Studies have suggested that AI can reduce interpretation time for radiologists [4] and even improve fracture detection sensitivity compared to junior radiologists [5]. However, these studies primarily assess AI's diagnostic capabilities, not its role in workflow changes – a critical gap in current research.

AI models used in fracture detection are trained on large datasets and often exhibit high sensitivity, meaning they excel at identifying potential fractures. However, this increased sensitivity comes at the expense of higher false positive rates, where AI mistakenly flags normal findings as fractures or experiences difficulty differentiating old fractures from new ones. While previous studies have examined false negative rates (i.e., AI missing fractures) as a safety concern, less attention has been given to false positives, which may lead to unnecessary imaging, overtreatment, and increased healthcare costs. The real-world implications of radiologists reassessing AI-positive cases post-treatment have not been systematically evaluated, making it essential to investigate whether this step improves patient outcomes or merely confirms decisions already made by orthopedic surgeons.

Workflow efficiency is a key priority for radiology departments as imaging volumes continue to rise and workforce shortages persist [6]. Reducing radiologist involvement in AI-positive cases could allow radiologists to focus on more diagnostically complex or uncertain cases, thereby improving overall patient care. However, any workflow modification must be empirically validated to ensure that it does not introduce unintended clinical risks. This study aims to address this by assessing whether post-treatment radiologist assessment of AI-positive cases adds clinical value. Unlike previous studies that primarily assess AI's diagnostic accuracy, this study focuses on workflow efficiency, resource utilization, and patient outcomes.

## III. Methodology and Ethical Consideration

### *A. Study design and setting*

This study is based on a retrospective, exploratory design, aimed at evaluating the clinical consequences of workflow modifications introduced by integrating AI-driven applications into clinical practice. The retrospective, exploratory study design is particularly suited for initial investigations of workflow modifications, as it allows real-world data to be analyzed without artificially altering standard practice [7]. The study is conducted in a hospital setting where approximately 2,300 patients undergo AI-supported radiological evaluation over a five-month period. The population includes patients for whom the BoneView application indicates a positive finding. Among these, cases where subsequent clinical evaluation determines no need for treatment – designated as false positives – are included in the final dataset.

The use of FP cases provides a focused and practical approach to investigating the potential added clinical value of radiologists' assessments in AI-positive workflows. These cases represent scenarios where BoneView identifies a positive finding, but subsequent human evaluation by physicians concludes with no relevant clinical findings. By examining such cases, the study isolates instances where radiologists' expertise might either confirm or challenge the AI's assessment. This enables the evaluation of whether radiologists' interpretations contribute to improved diagnostic accuracy, prevent overtreatment, or identify subtleties missed by AI alone.

False negative cases were not included in this study because they are not relevant to the proposed workflow modification. The study evaluates the impact of removing radiologists' assessment of AI-positive cases, meaning that all cases in which the AI indicates a negative finding will continue to be reviewed by radiologists as part of standard practice. Since false negatives occur when the AI fails to detect a fracture, these cases would still undergo radiologist evaluation under the modified workflow. As a result, their inclusion would not contribute to answering the primary research question, which focuses on whether post-procedure radiologist assessment of AI-positive cases adds clinical value. Focusing on FP cases allows for a targeted assessment and are particularly suited for this investigation since they highlight potential discrepancies in AI performance, offering critical insight into the role of human oversight in ensuring patient safety and diagnostic rigor.

### *B. Data collection and sampling*

FP cases will be identified prospectively at the point of care when either an orthopedic surgeon or a radiologist determines an AI-positive case to lack sufficient evidence of fracture. In these cases, patient identifiers will be recorded for retrospective analysis, see Figure 2 for study procedure. To ensure the rigor of the research, it is critical that the process of data collection remains independent of clinical decision-making and treatment. All patients will receive treatment-as-usual, following established clinical protocols, regardless of whether their case is identified and registered as FP. This independence ensures that the registration of FPs does not influence or alter the treatment decisions made by the clinicians, ensuring that the results accurately reflect the impact of workflow modifications without interference from the data collection process.

Based on expert estimates from the early implementation phase, the prevalence of FP is anticipated to be 1-2%. Based on this estimate, the study is expected to identify and include 20-40 FP cases during the five-month data collection period.

While this sample may be limited in detecting rare but high-impact clinical events, these cases represent a targeted subset derived from an estimated total of approximately 2,300 AI-evaluated X-ray examinations. The focused selection of FPs inherently constitutes a 'funnel' from a larger data pool, enhancing specificity and clinical relevance. Thus, despite the smaller number of cases included in detailed analyses, the extensive initial dataset bolsters the representativeness and applicability of our findings to broader clinical practice, and provides sufficient depth for an exploratory analysis of workflow effects.

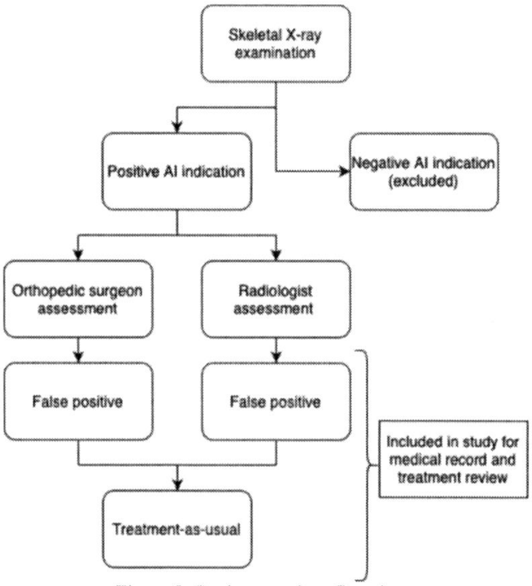


Figure 2. Study procedure flowchart.

## C. *Outcome measures and analysis*

The primary outcome measure is the potential clinical harm or unnecessary care resulting from false positive AI assessments when radiologist evaluation is omitted. Secondary outcomes include (1) the frequency of overtreatment related to AI-positive findings, (2) patient recall rate, (3) resource utilization metrics (e.g., additional imaging or extended clinical encounters), and (4) the concordance between orthopedic and radiologist evaluations. A panel of experienced clinicians (radiologists and orthopedic surgeons) will independently review the medical records and provided treatments of the included FP cases and apply structured criteria to evaluate the degree of clinical impact using a standardized rubric (see Table 1). To minimize potential biases introduced by differing clinician thresholds for identifying false positives, all included cases will undergo independent review by an expert panel comprising both radiologists and orthopedic surgeons. For each FP case, ground truth will be retrospectively set by a specialist in musculoskeletal radiology (i.e., the *"gold standard"*)

Using the documented cases, the study will analyze the clinical pathways to identify the role of radiologist's evaluations in mitigating adverse outcomes. The analysis will focus on mapping the potential effects of radiological input on treatment decisions, providing a qualitative and quantitative basis for assessing the workflow change. Quantitative data (e.g., number of FP cases, frequency of overtreatment) will be summarized using descriptive statistics. We will also conduct comparative analyses to explore relationships between FP cases and patient demographics or fracture types. Qualitative data, including expert panel evaluations of clinical impact, will be thematically analyzed using framework method [8] to identify patterns in decision-making and diagnostic discrepancies.

TABLE 1. INFORMATION TO BE COLLECTED DURING REVIEW PHASE

| Nr. | Category | Predefined items |
|---|---|---|
| 1 | Serial nr. | |
| 2 | PACS nr. | |
| 3 | Gender | Male; Female |
| 4 | Age | 0-20; 20-60; 60+ |
| 5 | Bone region | |
| 6 | Fracture type | |
| 7 | Other clinical findings | Bone lesion; Hydrops; Luxation |
| 8 | Patient referral from | |
| 9 | Patient sent to after X-ray examination | |
| 10 | AI fracture indication | Positive; Negative; Doubt |
| 11 | Image evaluation from radiologist | Positive; Negative; Doubt |
| 12 | Image evaluation from orthopedic surgeon | Positive; Negative; Doubt |
| 13 | Treatment implication | Overtreatment; Orthopedic FP identification; Radiologist FP identification; Patient recall |

Overtreatment is defined in this study as any unnecessary medical intervention (e.g., additional imaging, orthopedic procedures) that would not have occurred had the radiologist reviewed and correctly classified the AI-positive case. Patient recall refers to instances where a patient is asked to return for further evaluation due to an AI-generated false positive result. Resource utilization encompasses additional diagnostic procedures, prolonged clinical encounters, and increased workload for radiologists and orthopedic surgeons. AI-based systems like BoneView may occasionally flag incidental findings (e.g., bone lesions, luxations) that do not directly correspond to fractures. These cases are assessed by clinical reviewers to determine whether they contribute to false positive classifications and whether their presence influences resource utilization or overtreatment.

To translate expert chart reviews into meaningful metrics, we will apply a structured rubric that categorizes clinical consequences based on severity, distinguishing cases with minor clinical implications from those leading to significant clinical harm or unnecessary interventions.

## D. *Ethical considerations*

According to the Norwegian Act on Medical and Health Research §2 and §4, the study does not require approval from the regional ethics committee (REK). Data handling and storage will comply with institutional and national privacy standards and regulations, with approval from the hospital's Data Protection Officer. Informed consent will be obtained from patients identified as FP prior to accessing their medical records for detailed analysis. The study does not entail any change in workflow during the duration of the project, and all patients will receive treatment-as-usual.

## IV. EXPECTED OUTCOMES AND STUDY IMPLICATIONS

Randomized Controlled Trials (RCTs) are the gold standard for investigating clinical outcomes from clinical interventions. However, the number of studies adhering to this methodology that address clinical outcomes of AI implementations in healthcare are presently limited and concentrated around a few geographical clusters [9]. While clinical evidence for AI in healthcare is certainly in demand, we encourage an effort to develop study designs that can explore the clinical proxy outcomes of AI implementation in locations that have implemented AI in clinical practice. The outcomes of this study may provide critical insights into the integration of AI applications into clinical workflows, addressing a key gap in the existing knowledge [10]. While pre-market certifications validate the safety and clinical performance of AI applications, it does not address their operational downstream effects on clinical workflows and resource allocation. This study bridges this gap by exemplifying a study design to evaluate workflow modifications and their implications, addressing the growing need for post-deployment validation frameworks.

We expect the final findings to contribute to understanding how to optimize resource allocation without compromising patient safety. Specifically, the study aims to identify the extent to which radiological evaluations influence treatment decisions and mitigate risks in specific patient cases. Moreover, the study's findings could inform triage protocols by identifying which clinical cases warrant immediate radiologist input and which can safely proceed without additional radiologist review. By discerning patterns in AI performance – particularly for different fracture types or demographic groups – clinicians and administrators could prioritize high-risk patient cohorts for expedited evaluation. By specifically focusing on FP cases, this study isolates the cohort of patients most vulnerable to overtreatment and misdiagnosis in the absence of radiological review – thereby providing a high-fidelity assessment of the necessity for radiologist oversight in image assessment in non-complex facture cases.

This approach aligns with the broader literature emphasizing the need to balance efficiency and safety in healthcare AI deployment [11][12]. Additionally, the proposed methodology supports scalability and adaptability, offering a replicable method for other institutions to assess AI-induced workflow changes, and underscores the importance of clinician involvement to ensure that workflow adjustments maintain transparency and clinical relevance.

## REFERENCES


[1] A. Muley, P. Muzumdar, G. Kurian, and G. P. Basyal, "Risk of AI in Healthcare: A comprehensive literature review and study framework," *arXiv preprint arXiv:2309.14530,* 2023.

[2] L. Silsand, M. Kanneløning, G.-H. Severinsen, and G. Ellingsen, "Enabling AI in Radiology: Evaluation of an AI Deployment Process," *Studies in health technology and informatics,* vol. 316, pp. 580-584, 2024.

[3] R. Najjar, "Redefining radiology: a review of artificial intelligence integration in medical imaging," *Diagnostics,* vol. 13, no. 17, p. 2760, 2023.

[4] S. Jeong *et al.*, "The Impact of Artificial Intelligence on Radiologists' Reading Time in Bone Age Radiograph Assessment: A Preliminary Retrospective Observational Study," *Journal of Imaging Informatics in Medicine,* pp. 1-9, 2024.

[5] R. Y. Kuo *et al.*, "Artificial intelligence in fracture detection: a systematic review and meta-analysis," *Radiology,* vol. 304, no. 1, pp. 50-62, 2022.

[6] R. Bruls and R. Kwee, "Workload for radiologists during on-call hours: dramatic increase in the past 15 years," *Insights into imaging,* vol. 11, pp. 1-7, 2020.

[7] P. Aithal and S. Aithal, "Redefining Experimental, Empirical, and Exploratory Research in AI Era," *Poornaprajna International Journal of Emerging Technologies (PIJET),* vol. 1, no. 1, pp. 90-136, 2024.

[8] N. K. Gale, G. Heath, E. Cameron, S. Rashid, and S. Redwood, "Using the framework method for the analysis of qualitative data in multi-disciplinary health research," *BMC medical research methodology,* vol. 13, no. 1, pp. 1-8, 2013.

[9] T. Y. Lam, M. F. Cheung, Y. L. Munro, K. M. Lim, D. Shung, and J. J. Sung, "Randomized controlled trials of artificial intelligence in clinical practice: systematic review," *Journal of Medical Internet Research,* vol. 24, no. 8, p. e37188, 2022.

[10] A. Jackson and B. Hirsch, "Changing the workflow–Artificial intelligence in radiologic sciences," vol. 55, ed: Elsevier, 2024, p. 101710.

[11] M. A. Sujan, "Looking at the Safety of AI from a Systems Perspective: Two Healthcare Examples," in *Safety in the Digital Age: Sociotechnical Perspectives on Algorithms and Machine Learning*: Springer Nature Switzerland Cham, 2023, pp. 79-90.

[12] E. J. Topol, "High-performance medicine: the convergence of human and artificial intelligence," *Nature medicine,* vol. 25, no. 1, pp. 44-56, 2019.

# Using Explainable Machine Learning for Diabetes Management in Emergency Departments

Silas Majyambere
Department of Computer and Systems Sciences
Stockholm University (SU)
Stockholm, Sweden
E-mail: majyambere@dsv.su.se

Celestin Twizere
Center for Biomedical Engineering and E-Health (CEBE)
University of Rwanda (UR)
Kigali, Rwanda
E-mail: celestintwizere@gmail.com

Tony Lindgren
Department of Computer and Systems Sciences
Stockholm University (SU)
Stockholm, Sweden
E-mail: tony@dsv.su.se

Gerard Nyiringango
Center for Biomedical Engineering and E-Health (CEBE)
University of Rwanda (UR)
Kigali, Rwanda
E-mail: nyiringangogerard@gmail.com

***Abstract*—Uncontrolled diabetes can lead to severe complications and Intensive Care Unit (ICU) admissions. This study presents an explainable machine learning model using electronic health records to predict ICU admissions and estimate hospital stay duration for diabetic patients. AdaBoost model outperformed other models on ICU admission prediction, while CatBoost exhibited superior performance in estimating ICU length of stays among diabetic patients admitted to the emergency departments. The results demonstrate the potential of explainable machine learning in ICU risk assessment and can aid healthcare providers in early intervention and resource utilization. The clinician and the proposed model agree on the top 25 features identified by Shapley Additive exPlanations (SHAP) methods for predicting ICU admission, but they differ in the ranking of the top five most significant predictors.**

***Keywords-Explainable Machine Learning; Intensive Care Unit; Diabetes; Length of Stay; SHAP.***

## I. INTRODUCTION

Diabetes is a chronic disease affecting millions globally and is a major contributor to morbidity and mortality. There are three kinds of diabetes: Type 1 Diabetes Mellitus (T1DM), which results from the pancreas's inability to produce insulin; Type 2 Diabetes Mellitus (T2DM), characterized by the body's ineffective use of insulin [1][2]; and gestational diabetes, which occurs during pregnancy and may later be resolved or progress to T1DM or T2DM. Despite medical advancements, diabetes prevalence continues to rise, with cases estimated at 536 million in 2021 and projected to reach 783 million by 2045 [3]. Managing diabetes remains challenging due to complications, such as cardiovascular diseases, neuropathy, nephropathy, retinopathy, and glycemic complications, often leading to frequent Emergency Department (ED) visits and Intensive Care Unit (ICU) admissions [4]. This study focuses on T1DM and T2DM.

Emergency departments play a crucial role in managing critically ill diabetic patients and serve as valuable data sources for predicting ICU admissions and estimating hospital stays. Machine Learning (ML) models trained on ED data can assist clinicians in identifying high-risk patients in real-time, allowing for timely interventions. In the medical domain, Explainable Artificial Intelligence (XAI) techniques, such as SHAP, are essential to ensure the reliability and clinical applicability of AI-enabled tools. XAI fosters trust among ED healthcare providers by ensuring that ML models use relevant and medically validated features. Additionally, XAI provides valuable insights into critical risk factors, improving decision-making in emergency care settings.

Machine learning has been widely applied during the COVID-19 pandemic to predict ICU admissions [5], as well as to estimate ICU length of stay [6] and readmission risk among diabetic patients [7]. However, the application of explainable ML in emergency settings for diabetes management remains underexplored. The contribution of this study is in three folds: (1) Develop boosted tree-based ensemble machine learning models using ED data to predict ICU admission risk for T1DM and T2DM patients, (2) Build a machine learning model to estimate the length of hospital stays for diabetes patients upon ED admission, (3) Apply SHAP methods to provide interpretable explanations for the classification and regression models predicting ICU admissions and length of stay.

This study aims to develop explainable ML models using boosted tree ensemble algorithms and SHAP to predict ICU admission risk and estimate ICU length of stay for diabetic patients in the ED. The dataset for this study includes historical patient data, including demographics, vital signs, and lab results from electronic health records stored in the Medical Information Mart for Intensive Care (MIMIC)-IV database [19]. Performance evaluation metrics used for ICU admission prediction (classification task) are accuracy, precision, recall, F1-score, and Area Under the Curve score (AUC), while hospital stay estimation (regression task) was evaluated using Root Mean Squared Error (RMSE) and Mean Absolute Error (MAE) metrics.

The remainder of this paper is structured as follows: In Section 2, we outline the methodology guiding our study, Section 3 presents the experimental results, Section 4 provides a discussion of the findings, and Section 5 concludes the study while outlining directions for future research.

## II. METHODOLOGY

This study has developed ML models to predict diabetes ICU admissions and length of stay using MIMIC-IV data and five machine learning models, including Decision Tree, AdaBoost, CatBoost, XGBoost, and LightGBM. These boosted tree ensemble models were chosen for their superior performance on ICU and Length Of Stays (LOS) prediction, as reported in the literature [16][17][18]. Additionally, they effectively address class imbalance, making them well-suited for this study. Performance was evaluated using classification and regression metrics. SHAP was applied to ensure interpretability and clinical relevance. This study was conducted in six steps: data acquisition, dataset building, data preprocessing, model building, performance evaluation, and generating explanations of the best model, as shown in Figure 1.

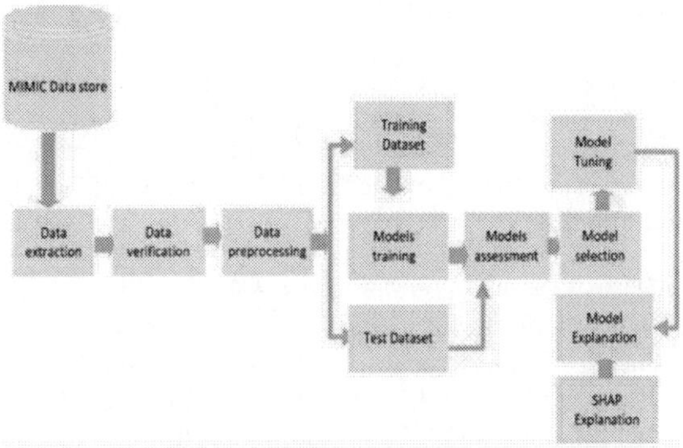


Figure 1. Design of the research processes.

### A. *Data acquisition*

We used MIMIC-IV (version 2) [19], focusing on MIMICIV-ED, MIMICIV-ICU, and MIMICIV-Hospital datasets, extracted via BigQuery SQL. Diabetic patients were identified using the International Classification of Diseases (ICD)-10 codes: E10.XXX for T1DM and E11.XXX for T2DM [20]. MIMIC-IV extends MIMIC-III, providing electronic health records for Beth Israel Medical Center ICU patients from 2008 to 2019. The publicly available database contains deidentified patient data in compliance with the Health Insurance Portability and Accountability Act (HIPAA). Figure 2 highlights the tables and the type of data extracted from the MIMIC-IV database.

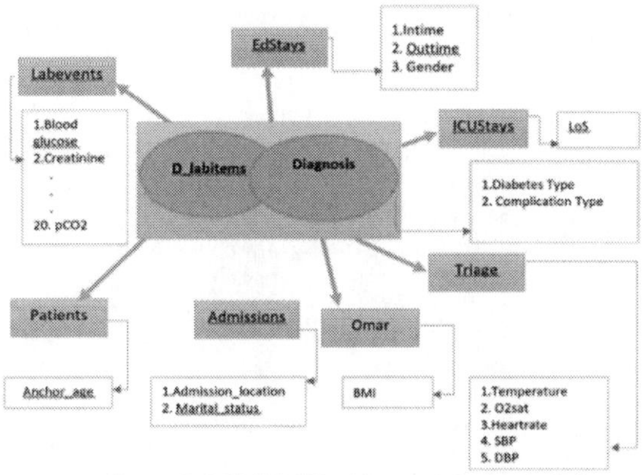


Figure 2. MIMIC-IV tables used dataset building.

Data was extracted using ICD-10 codes. Subject ID represents individual patients, while item ID corresponds to lab tests during hospitalization. More data was extracted from the Labevents table (20 features).

### B. *Dataset building*

Between 2008 and 2019, 9339 diabetic patients were hospitalized, including 1090 with T1DM and 7097 with T2DM. Focusing on first-time emergency admissions, 4365 patients were selected after removing duplicates (863 T1DM, 3502 T2DM). The study includes demographic data, vital signs, diagnoses, and lab results for critically ill diabetic patients. We extracted 34 numerical features (detailed in Table I) and 6 categorical features, resulting in 20 features after label and one-hot encoding, as illustrated in Figure 3.

### C. *Data preprocessing*

Data preprocessing is a crucial step in machine learning, involving missing data handling, feature correlation analysis, deduplication, outlier removal, and data scaling to enhance model performance. Categorical variables were encoded using label encoding for binary features and one-hot encoding for multi-value features. To maintain a focus on T1DM and T2DM, cases related to pregnancy, malnutrition, and unspecified causes were excluded, reducing the dataset to 4,317 patients. Outlier detection was performed by analyzing each feature individually for anomalies due to human error or lack of relevance to the study. Extreme deviations from expected distributions, such as a Body Mass Index (BMI) of 3,658.50, systolic blood pressure of 1.00, or a temperature of 9°F (as shown in Table I), were considered outliers and removed. Features with more than 50% missing values, including Protein C (PC), Protein Creatinine Ratio (PCR), and Hemato, were removed to prevent bias to the ML model. Finally, two datasets were created: one for ICU admission prediction (4,055 samples, 49 features) and another for ICU length of stay prediction, patients with a zero-day ICU stay were excluded from ICU LOS dataset, as they were not admitted leading to a dataset of 1,432 samples, and 49 features.

TABLE I. DESCRIPTIVE STATISTICS

| Feature Name | Count | Min | Max | Mean | Missing Values (%) |
|---|---|---|---|---|---|
| Temp | 3721.00 | 9.00 | 106.30 | 97.89 | 13.81 |
| Hrate | 3776.00 | 20.00 | 220.00 | 84.97 | 12.53 |
| Resprate | 3758.00 | 8.00 | 40.00 | 17.74 | 12.95 |
| O2sat | 3737.00 | 60.00 | 109.00 | 98.23 | 13.44 |
| SBP | 3687.00 | 1.00 | 249.00 | 143.64 | 14.59 |
| DBP | 3633.00 | 14.00 | 474.00 | 78.15 | 15.84 |
| Age | 4317.00 | 18.00 | 91.00 | 58.53 | 0.00 |
| BMI | 3964.00 | 0.67 | 3658.5 | 54.83 | 8.18 |
| Alb | 3765.00 | 1.20 | 5.85 | 3.81 | 12.79 |
| Hemog | 4148.00 | 3.20 | 19.60 | 11.64 | 3.91 |
| Cr | 4287.00 | 0.10 | 16.30 | 1.66 | 0.69 |
| BG | 4295.00 | 28.00 | 1083.00 | 200.01 | 0.51 |
| Trig | 3061.00 | 10.00 | 7140.00 | 189.43 | 29.09 |
| HbA1C | 3495.00 | 4.00 | 19.40 | 8.24 | 19.04 |
| RBC | 4285.00 | 1.77 | 6.65 | 4.08 | 0.74 |
| WBC | 4216.00 | 0.20 | 181.70 | 10.83 | 2.34 |
| Sod | 4275.00 | 82.00 | 177.00 | 136.72 | 0.97 |
| Pot | 4252.00 | 1.80 | 10.00 | 4.78 | 1.51 |
| pH | 2465.00 | 6.64 | 7.86 | 7.34 | 42.90 |
| pO2 | 2395.00 | 12.00 | 536.00 | 120.34 | 44.52 |
| pCO2 | 2387.00 | 8.00 | 199.00 | 40.25 | 44.71 |
| LDL | 2938.00 | 10.00 | 407.00 | 96.13 | 31.94 |
| HDL | 3017.00 | 2.10 | 214.00 | 48.35 | 30.11 |
| EdLOS | 4317.00 | 0.00 | 4.73 | 0.55 | 0.00 |
| ICUStays | 1553.00 | 0.00 | 50.33 | 2.76 | 64.03 |
| BUN | 4286.00 | 1.00 | 260.00 | 23.60 | 0.72 |
| Bilir | 3456.00 | 0.10 | 29.20 | 0.82 | 19.94 |
| PC | 1999.00 | 6.00 | 1697.00 | 250.28 | 53.69 |
| CholR | 3027.00 | 1.30 | 36.00 | 3.97 | 29.88 |
| CK | 3049.00 | 8.00 | 68510.00 | 332.95 | 29.37 |
| VB12 | 2337.00 | 88.00 | 1976.00 | 671.05 | 45.87 |
| Iron | 2494.00 | 8.00 | 396.00 | 61.08 | 42.23 |
| Hemato | 1286.00 | 10.00 | 60.00 | 32.76 | 70.21 |
| PCR | 1065.00 | 0.10 | 355.50 | 1.86 | 75.33 |

*Abbreviations:* Temp: Temperature, Hrate: Heart rate, Resprate: Respiration rate, O2sat: Oxygen saturation, SBP: Systolic blood pressure, DBP: Diastolic blood pressure, Alb: Albumin, Hemog: Hemoglobin, Cr: Creatinine, BG: Blood glucose, Trig: Triglycerides, HbA1C: Glycated Hemoglobin, RBC: Red Blood Cells, WBC: White Blood Cells, Sod: Sodium, Pot: Potassium, pO2: Partial pressure of oxygen, pCO2: Partial pressure of Carbon Dioxide, LDL: Low-Density Lipoprotein sometimes called bad cholesterol, HDL: High-Density Lipoprotein known as good cholesterol, BUN: Blood Urea Nitrogen, Bilir: Bilirubin (Total), CholR: Cholesterol ratio, CK: Creatinine kinase, VB12: Vitamin B12.

Our experiment shows the distribution of categorical features (Figure 3) in the dataset, which includes 2293 females and 2024 males, with 3464 type 2 diabetes and 853 type 1 diabetes patients. A total of 4317 T1DM and T2DM patients were admitted to the emergency department, with 1560 admitted to the ICU, indicating not a fully balanced dataset. The "Admit" feature has four categories for admission status, while the "CompType" feature includes nine diabetes-related complications. The "Marital Status" feature has four categories: married, widowed, single, and divorced. Widowed, single, and divorced are often grouped as single, but they are treated separately here due to psychological differences that may affect diabetes differently.

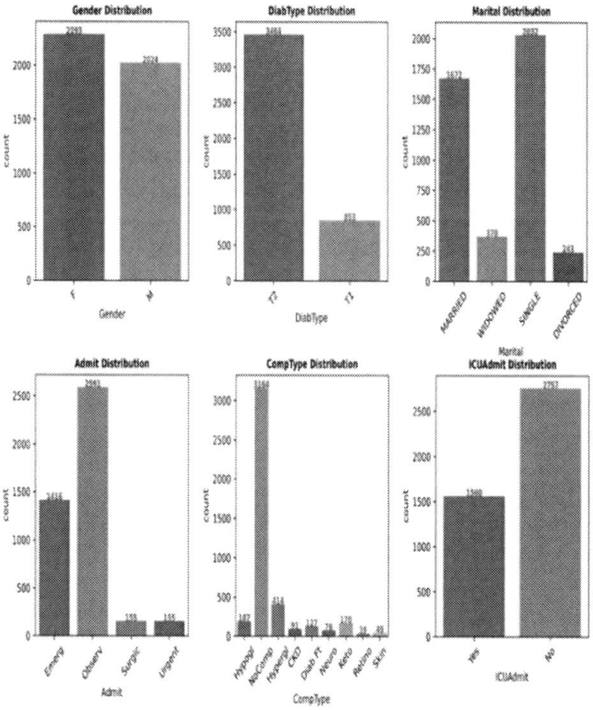

Figure 3. Visualizing the distribution of categorical data.

We used the Robust scaler method to minimize the effect of features with very low or very high values to put all numerical data on the same scale. The formula of a robust scaler is:

$$X = \frac{X - X_{median}}{IQR} \qquad (1)$$

A robust scaler is less affected by outliers. IQR means the InterQuartile Range between the first quartile (25%) and the third quartile (75%).

### D. *Machine Learning Algorithms*

This study uses supervised machine learning with five models: Decision Tree [8], CatBoost [9], AdaBoost [10], XGBoost [12], and LightGBM [13] for ICU admission prediction and length of stay estimation. Initial experiments involve training these models with default parameters on cleaned datasets. The best model was selected and fine-tuned through hyperparameter optimization to improve predictive performance.

#### (1) *Decision Tree Algorithm*

A decision tree model is a simple predictive modeling tool that can be used for classification and regression tasks. It works by building a tree where the nodes are the decision rules, and the leaf nodes give the output of the prediction [8]. This type of model is known as the "white model" since it is easier to understand and interpret its prediction process. The decision tree model is affected by the curse of dimensionality, where a large number of features will increase the splitting process, which results in poor performance of the model.

Tree-based ensemble models were introduced to solve the limitations of decision tree models operating on large datasets.

(2) *CatBoost Algorithm*

CatBoost is an open-source gradient-boosted tree library that excels in handling categorical data without the need for transformation. Developed by Yandex, CatBoost is well-suited for machine learning tasks involving heterogeneous data, particularly when categorical features are present. It builds a model iteratively through gradient boosting, improving its performance step by step using weak learners, typically decision trees [9]. One of CatBoost's key advantages is its ability to handle categorical data efficiently, reducing the need for pre-processing transformations often required by other machine learning models.

(3) *AdaBoost Algorithm*

AdaBoost is a boosted tree ensemble algorithm that improves the weights of weak learners over multiple iterations to make predictions. It builds a strong classifier by training weak classifiers sequentially, with more emphasis on misclassified instances in each iteration. AdaBoost is particularly effective for binary classification and regression tasks [10]. Its drawback is that training weak learners sequentially leads to longer training time when the dataset is large.

(4) *XGBoost Algorithm*

XGBoost (Extreme Gradient Boosting) builds trees sequentially, with each tree correcting the errors of the previous ones through gradient optimization. It is an open-source library that uses distributed gradient-boosted trees for predictions. XGBoost combines regularization and optimization techniques to increase predictive performance and reduce the training time. Known for its high prediction accuracy and fast training speed, XGBoost is widely used in classification and regression tasks with structured data [11] [12]. The base model used in XGBoost is the Classification And Regression Tree (CART).

(5) *LightGBM Algorithm*

LightGBM is a gradient-boosting library developed by Microsoft for classification and regression tasks. It builds decision trees using a leaf-wise approach, enabling faster training. Its strong performance is driven by two key innovations: Gradient-Based One-Side Sampling (GOSS), which prioritizes informative data points by removing small-gradient samples, and Exclusive Feature Bundling (EFB), which reduces dimensionality by combining mutually exclusive features. Additionally, LightGBM addresses class imbalance by assigning higher weights to the minority class [13].

E. *Experiment Setup*

We used Python and its libraries, such as Pandas, NumPy, Matplotlib, Seaborn, and Scikit-learn, and the SHAP library for explainability and other libraries required for this study. The experiment began with exploratory data analysis, followed by feature engineering to prepare the final datasets. In collaboration with a clinician, we selected relevant features for use in the chosen machine learning models. Significant effort was dedicated to data preprocessing to ensure the creation of a high-quality dataset suitable for both ICU admission prediction and ICU length of stay estimation.

F. *Training and Evaluating Models*

We assessed the performance of the proposed model using standard evaluation metrics and analyzed the interpretability of its predictions through SHAP-based explanations

(1) *Performance metrics*

The selected machine learning models support both classification and regression tasks. In both cases, models were trained on 80% of the data and validated on the remaining 20%. For the classification task (ICU admission prediction), performance was evaluated using accuracy, precision, recall, F1-score, and AUC according to equations (2–5). The Receiver Operating Characteristics (ROC)-AUC curve was used for model comparison, and the best-performing model underwent hyperparameter tuning with GridSearchCV and internal validation via K-Fold cross-validation with K equals to 10. The optimized model was then used for final predictions, with its learning curve analyzed and feature importance compared to SHAP's global explanations. For the regression task (ICU length-of-stay estimation), models were evaluated using RMSE and MAE. The best model's hyperparameters were tuned before predicting ICU stays for diabetic patients. SHAP was applied to provide local explanations, illustrating why specific predictions were made.

$$Accuracy = \frac{TP + TN}{TP + TN + FP + FN} \quad (2)$$

$$Precision = \frac{TP}{TP + FP} \quad (3)$$

$$Recall = \frac{TP}{TP + FN} \quad (4)$$

$$F1 - score = \frac{2*TP}{2*TP + FP + FN} \quad (5)$$

AUC and ROC-AUC curves are calculated using the probability distribution of True Positive Rate (TPR) and False Positive Rate (FPR), as indicated in Figure 5. The regression models used to estimate the LOS, were assessed using RMSE and MAE calculated using the below formula:

$$RMSE = \sqrt{\frac{1}{n}\sum_{i=1}^{n} (y_i - y_p)^2} \quad (6)$$

$$MAE = \frac{1}{n}\sum_{i=1}^{n} |y_i - y_p| \quad (7)$$

where $n$ is the total number of samples (rows), $y_i$ is the true value and $y_p$ is the predicted value.

*(2) Model Explanations using SHAP Methods*

SHAP is a widely used post-model explanation method in XAI, using cooperative game theory to compute Shapley values that quantify each feature's contribution to model predictions. It is characterized by four key properties: Efficiency, Symmetry, Dummy, and Additivity, as outlined by Lundberg and Lee [14]. Shapley values are calculated using the following formula:

$$\phi_i(f) = \sum_{S\subseteq N\setminus i} \frac{|s|!(|N|-|S|-1)!}{N!}\left[f(S \cup i) - f(s)\right] \quad (8)$$

Here, $S$ is a subset of features excluding feature $i$, and $N$ is the total number of features. $\phi_i(f)$ represents the contribution of feature $i$, calculated as the difference between the total prediction and the prediction without feature $i$.

Due to their computational complexity, Shapley values can be expensive to compute. To address this, we used TreeSHAP [15], an optimized version of SHAP for tree-based ensemble models. TreeSHAP employs path-dependent feature perturbation, reducing computational complexity to $O(TLD^2)$ in the worst case, compared to the original SHAP's complexity of $O(2^n)$, where $n$ is the number of features, $T$ is the number of decision trees, $L$ is the number of leaves, and $D$ is the depth.

## III. EXPERIMENTAL RESULTS

This section presents findings on ICU admission prediction and length-of-stay estimation for diabetic patients visiting the emergency department.

### A. *ICU Admission Prediction*

We illustrate feature correlations with ICU admission on Figure 4, highlighting the top ten predictors: Albumin (Alb), Hemoglobin (Hemog), Creatinine (Cr), Red Blood Cell count (RBC), White Blood Cell count (WBC), Blood Urea Nitrogen (BUN), Total Bilirubin (Bilir), pH, Emergency (Emerg), and Observation (Observ).

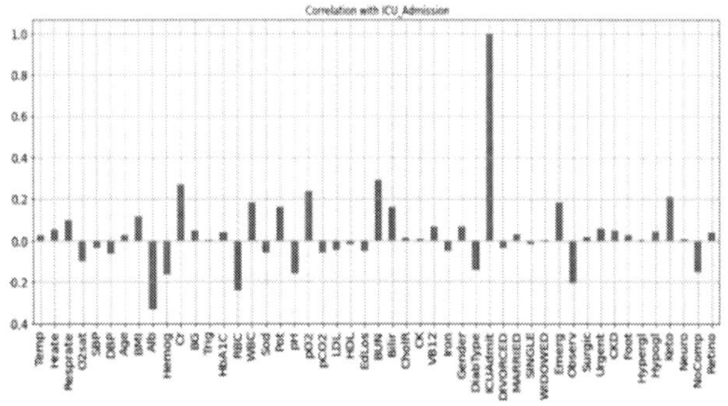


Figure 4. Features correlation to the ICU admission.

The five machine-learning models were trained on 80% of the data and tested on the remaining 20%. The experimental results demonstrate that the selected ensemble models effectively identify diabetic patients at risk of ICU admission, with all boosted tree models achieving an AUC above 0.85 (Figure. 5). CatBoost outperformed other models on the AUC metric. The dashed yellow line indicates the random guess, and its AUC is 0.5. All selected models performed well on ICU admission prediction as indicated in Table II. AdaBoost excels in most evaluation metrics.

TABLE II. MODEL PERFORMANCE ON ICU ADMISSION PREDICTION

| Classifier | Accuracy | Precision | Recall | F1-score | AUC |
|---|---|---|---|---|---|
| Decision Tree | 74.60 | 66.33 | 64.57 | 65.44 | 0.726 |
| XGBoost | 80.76 | 77.24 | 68.54 | 72.63 | 0.861 |
| CatBoost | 82.24 | 79.04 | 71.19 | 74.91 | 0.882 |
| LightGBM | 81.75 | 78.73 | 69.87 | 74.04 | 0.869 |
| AdaBoost | 82.74 | 80.92 | 70.20 | 75.18 | 0.876 |

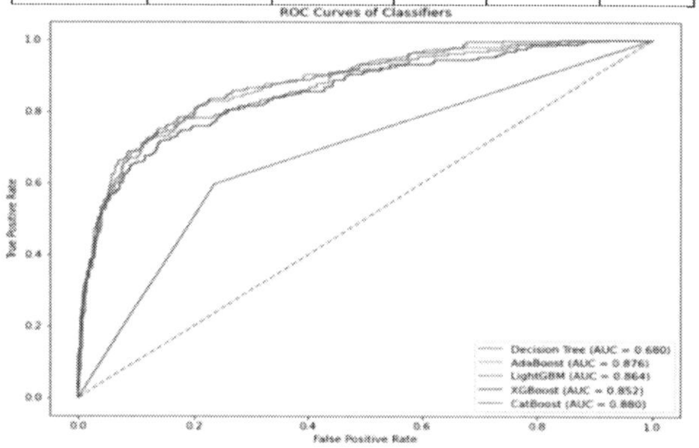


Figure 5. ROC Curve of ML models used to predict ICU admission with AUC between 0 and 1.

The learning curve shows that training and validation accuracy stabilize after 3,500 samples (Figure 6), indicating good generalization. This suggests the model reliably predicts ICU admissions.

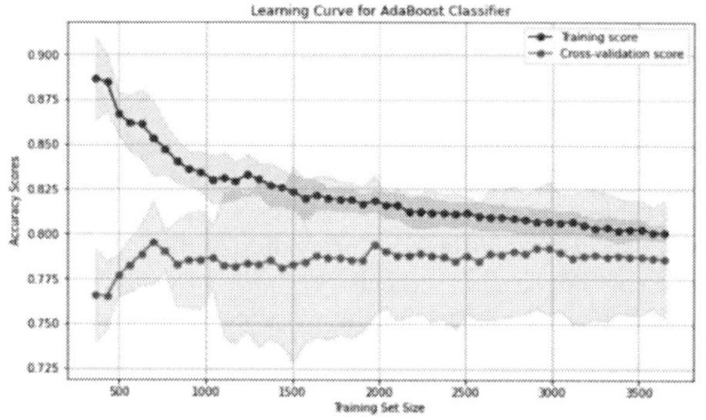


Figure 6. Accuracy of AdaBoost learning curve.

Using the internal architecture of AdaBoost and SHAP methods, we identified 25 most influential features for patients at risk of ICU admission, as shown in Figure 7.

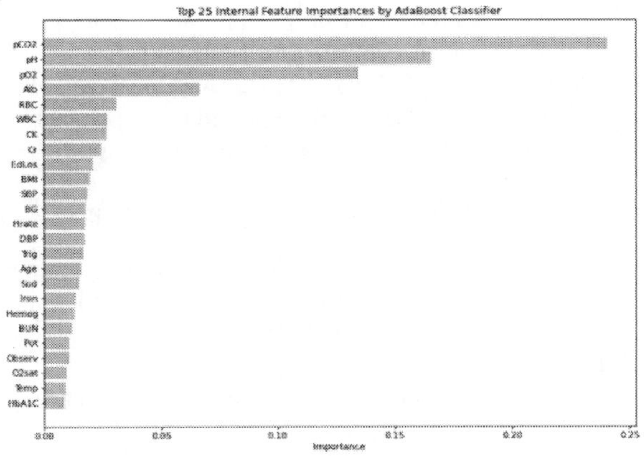

Figure 7. (a) Top 25 predictors by AdaBoost.

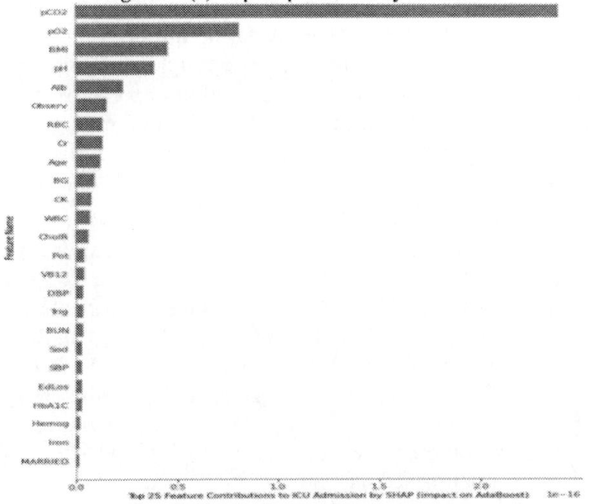

Figure 7. (b) Top 25 predictors by SHAP.

AdaBoost and SHAP identified the same four out of top five predictors of ICU admission in diabetic patients: pCO2, pH, $pO_2$, and Alb. The top 25 features in both AdaBoost and SHAP differed on five features: AdaBoost included O2sat and Temp, while SHAP added CholR, VB12, and MARRIED. pCO2 was further assessed as it is the top predictor of ICU admission (normal range is 35 – 45 mmHg). We found that most patients admitted to the ICU have pCO2 beyond the normal range.

We further trained the AdaBoost model using the top 25 predictors identified by SHAP. As shown in Table III, the model achieved performance comparable to that trained on the full feature set, demonstrating the effectiveness of SHAP in identifying key risk factors for ICU admission.

TABLE III. ADABOOST PREDICTIVE PERFORMANCE USING FEATURES IDENTIFIED BY THE SHAP METHOD.

| Classifier | Accuracy | Precision | Recall | F1-score | AUC |
|---|---|---|---|---|---|
| AdaBoost | 82.74 | 77.01 | 73.26 | 75.09 | 0.881 |

The clinician and SHAP agree on 25 key features associated with ICU admission, though they differ in their relevance rankings. A comparison of their rankings of top five features is summarized in Table V. Importantly, both agree on three of the top five features identified by SHAP.

TABLE IV. ANALYSIS OF SHAP EXPLANATIONS BY CLINICIAN

| No. | Model's top five | Clinician's top 5 | Agreement |
|---|---|---|---|
| 1 | pCO2 | BG | No |
| 2 | pO2 | pO2 | Yes |
| 3 | BMI | BMI | Yes |
| 4 | pH | Cr | No |
| 5 | Alb | Alb | Yes |

*B. Length Of Stays (LOS) Prediction*

We have used five regression ML models to estimate the length of stays among diabetic patients admitted to the hospital with the status of emergency. Table V shows the performance of the selected models on RMSE and MAE evaluation metrics. CatBoost model outperformed the other models with 2.454 and 1.695 days in ICU for RMSE and MAE on prediction of ICU LOS for diabetic patients admitted in the emergency department.

TABLE V. RSME AND MAE FOR FIVE REGRESSION MODELS WHILE ESTIMATING ICU LOS.

| Regression Model | RMSE | MAE |
|---|---|---|
| CatBoost | 2.454 | 1.695 |
| LightGBM | 2.739 | 1.937 |
| XGBoost | 2.863 | 1.951 |
| Decision Tree | 4.547 | 2.521 |
| AdaBoost | 4.864 | 4.537 |

Using SHAP, we visualized the explanations of predicted LOS of 2.62 days in ICU for a diabetic patient, as indicated by $f(x)$ on Figure 8. The ground truth is that this patient was previously hospitalized for 3.10 days in ICU.

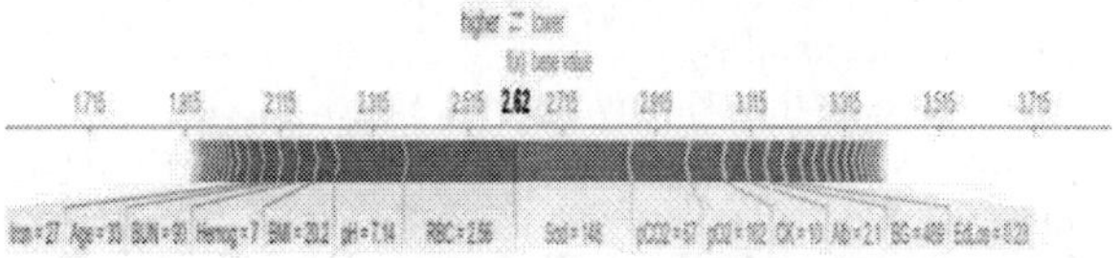

Figure 8. The SHAP force plot explaining ICU LOS for the patient on row number 51 in the test set.

SHAP was used to give details as to why this patient was predicted to stay in the ICU for 3.79 days in Figure 9. With the help of the highlighted features, we can see that features with Shapley values in red color will increase the number of days in the ICU, and features in blue will lower ICU stays.

This patient has a blood glucose of 264 and an HbA1C of 13.6, indicating poor control of blood sugar levels. In addition, the patient has pCO2 of 22 (very low) and Ketoacidosis complication, which is a major cause of prolonged stays in ICU among diabetic patients.

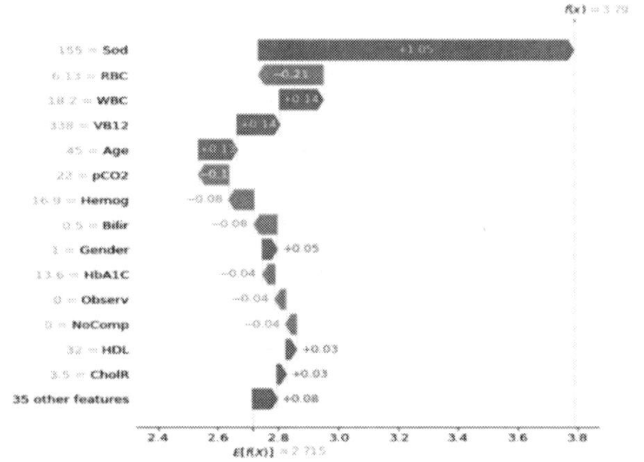


Figure 9. SHAP waterfall plot for estimating ICU stays for a patient with row 241 in the test set.

## IV. Discussion

In this retrospective study, we employed supervised machine learning to predict ICU admission among diabetic patients in the emergency department and estimate their ICU Length Of Stays (LOS). We evaluated four boosted tree ensembles and one decision tree model for classification and regression tasks. Despite the class imbalance, the models performed well, with AdaBoost achieving superior performance for ICU admission prediction (Table II) and CatBoost outperforming other models in estimating ICU LOS (Table V).

AdaBoost was selected for further optimization in ICU admission prediction and fine-tuned using GridSearchCV with 10-fold cross-validation. The best hyperparameters for classification were base_estimator = DecisionTreeClassifier (max_depth=5), learning_rate = 0.005, and n_estimators = 300. For regression, CatBoost was optimized with learning_rate = 0.05, n_estimators = 100, and max_depth = 8. Model predictions were interpreted using SHAP to enhance explainability.

To validate the effectiveness of SHAP explanations, we retrained AdaBoost using only the top 25 SHAP-selected features and achieved similar performance (Table III). For LOS estimation, CatBoostRegressor outperformed other models. SHAP explanations demonstrated clinical relevance, helping interpret model decisions. Figure 9 illustrates a case where a female patient with WBC = 18.2 (above the normal range: 4.5-11), Sod = 155 (above the normal 135–145 range), Age = 45, and VB12 = 338 (150-399 pg/mL indicates low levels of VB12) had a predicted ICU stay of over 3.5 days. These predictors, alongside abnormal values for BG and HbA1C, are clinically significant, as they are associated with uncontrolled blood glucose, which is a primary cause of diabetic emergency admissions at the hospital. To ensure the relevancy of our study, a medical practitioner specializing in diabetes management confirmed that our model is fair and accurately identifies key risk factors associated with ICU admission and extended length of stay for diabetic patients and suggested its implementation in the clinical workflow.

## V. Conclusion And Future Work

This study explored an explainable machine learning approach for diabetes management in emergency departments, focusing on early ICU admission prediction and length-of-stay estimation. Among five classifiers, the AdaBoost model demonstrated superior performance on three out of five metrics, as shown in Table II. CatBoost outperformed other tree-based models in regression tasks in predicting ICU stay duration as shown in Table V. SHAP analysis provided interpretability for both tasks, reinforcing the model's reliability. Our findings highlight the potential of ML integration in clinical workflows, most importantly in the emergency department since all critically ill patients start in this hospital unit, improving early ICU risk identification, optimizing hospital resource utilization, and enhancing emergency care for diabetic patients.

The experimental results successfully addressed the study's objectives: developing predictive models for ICU admission and length of stay among diabetic patients and generating interpretable explanations using SHAP. AdaBoost achieved the best performance for ICU admission prediction, while CatBoost excelled in estimating ICU length of stay. SHAP methods revealed the top 25 influential features for ICU admission prediction. A clinician with over 15 years of experience confirmed agreement with the features identified by SHAP, with minor differences in top feature rankings. Notably, the clinician appreciated that marital status was ranked lowest, aligning with its minimal clinical relevance in emergency diabetes care.

Considering the agreement and disagreement between the proposed model and clinical judgment in the Table IV, we recommend that emergency departments prioritize laboratory examinations of $pCO_2$, $pO_2$, BMI, Blood Glucose (BG), Creatinine (Cr), Albumin (Alb), and pH levels for diabetic patients presenting in emergency situations. This prioritization may enhance the accurate identification of patients at risk of ICU admission.

In future work, we aim to integrate the pretrained models into a Web application for real-world deployment in emergency departments. This will enable assessment of the model's effectiveness in identifying diabetic patients at risk of ICU admission. We also plan to evaluate the correlation between predicted and actual ICU length of stay. Finally, we will assess the impact on diabetes emergency care and the level of trust among diabetes care providers.

## ETHICAL APPROVAL

MIMIC-IV is a publicly available dataset accessed through PhysioNet and does not require additional ethical approval. However, to ensure patient anonymity, identifiable features such as subject ID and item ID, which were used to extract diabetic patient data, were removed.

## ACKNOWLEDGMENT

The authors express their gratitude to the University of Rwanda and Stockholm University for their support and facilitation during this study. This research was funded by the UR-Sweden Program for Research and Capacity Building Program through a scholarship to Silas Majyambere. All authors contributed equally to writing the manuscript. Silas was responsible for research design, data extraction, experiments, and drafting the write-up. Tony and Celestin supervised the study, enhanced the manuscript, and analyzed the experimental results. Gerard, leveraging his medical expertise in diabetes management for over 15 years, assisted in selecting relevant features for the dataset and assessing the clinical significance of the findings of the study.

## REFERENCES


[1] A. Oyeranmi, F. Okikiola, N. Yekini, and B. Ronke, "Classification of Diabetes Types using Machine Learning", IJACSA, vol. 13, No. 9, pp. 152-161, 2022.

[2] L. Kopitar, P. Kocbek, L. Cilar, A. Sheikh, and G. Stiglic, "Early detection of type 2 diabetes mellitus using machine learning-based prediction models". Nature Scientific Reports, 2020. https://doi.org/10.1038/s41598-020-68771-z.

[3] IDF Diabetes Atlas, "International Diabetes Federation", 10th edn. Brussels, 2021, https://www.diabetesatlas.org.

[4] Y. Jian, M. Pasquier, A. Sagahyroon, and F. Aloul, "A Machine Learning Approach to Predicting Diabetes Complications", MDPI Healthcare, 2021, https://doi.org/10.3390/healthcare9121712

[5] S. Saadatmand, K. Salimifard K, R. Mohammadi, A. Kuiper, M. Marzban, and A. Farhadi, "Using machine learning in prediction of ICU admission, mortality, and length of stay in the early stage of admission of COVID-19 patients", Annals of Operation Research, 2022. doi: 10.1007/s10479-022-04984-x.

[6] Y. Hu, L. Zheng, and J. Wang, "Predicting ICU Length of Stay for Patients with Diabetes Using Machine Learning Techniques", ICCSI, pp. 417-422, 2022.

[7] T. L. Hu, C. M. Chao, C. C. Wu, T. N. Chien, and C. Li, "Machine Learning-Based Predictions of Mortality and Readmission in Type 2 Diabetes Patients in the ICU", MDPI Applied Sciences, vol. 14, issue 18, 2024, https://doi.org/10.3390/app14188443.

[8] M. D. Maniruzzaman, M. D. J. Rahman, A. Benojir, and M. D. A. Menhazul, "Classification and prediction of diabetes disease using machine learning paradigm", HISS, vol. 8, 2020, https://doi.org/10.1007/s13755-019-0095-z.

[9] H. E. Massari, N. Gherabi, F. Qanouni, and S. Mhammedi, "Diabetes Prediction Using Machine Learning with Feature Engineering and Hyperparameter Tuning", IJACSA, vol. 15, no.8, 2024.

[10] X. Dong, K. Yu, and Z. Cui, "Readmission prediction of diabetic patients based on AdaBoost-RandomForest mixed model", ICBAIE, pp. 130-134, 2022.

[11] M. R. Khurshid, S. Manzoor, T. Sadiq, L. Hussain, M. S. Khan, and A. K. Dutta, "Unveiling diabetes onset: Optimized XGBoost with Bayesian optimization for enhanced prediction". PLOS ONE, vol. 20, 2025. doi.org/10.1371/journal.pone.0310218

[12] M. Liu, C. Guo, and G. Sijia, "An explainable knowledge distillation method with XGBoost for ICU mortality prediction", Computers in Biology and Medicine, vol. 152, 2023. https://doi.org/10.1016/j.compbiomed.2022.106466.

[13] G. Ke et al., "Lightgbm: A highly efficient gradient boosting decision tree". NIPS, pp. 3149 - 3157, 2017.

[14] S. M. Lundberg and S. Lee, "A Unified Approach to Interpreting Model Predictions", NIPS, 2017.

[15] S. M. Lundberg et al., "From local explanations to global understanding with explainable AI for trees". Journal of NMI 2, pp. 56–67, 2020. https://doi.org/10.1038/s42256-019-0138-9.

[16] L. Yuxi and Q. Shaowen, "An Interpretable Machine Learning Approach for Predicting Hospital Length of Stay and Readmission", ADMA 17th International Conference, pp. 73–85, 2022. https://doi.org/10.1007/978-3-030-95405-5_6.

[17] A. Gandra, "Predicting Hospital Readmissions in Diabetes Patients: A Comparative Study of Machine Learning Models", 2024. IJHS, vol. 8, no.3, pp. 289-297, 2024. https://doi.org/10.53730/ijhs.v8n3.15189.

[18] D. M. Somya, L. Deborah, T. P. McCoy, M. Jenkins, and M. Prashanti, "Machine learning for predicting readmission risk among the frail: Explainable AI for healthcare", Patterns, vol. 3, 2021. doi:10.1016/j.patter.2021.100395.

[19] E. W. J. Alistair et al., "Author Correction: MIMIC-IV, a freely accessible electronic health record dataset." Scientific data, vol. 10, 2023. doi:10.1038/s41597-023-01945-2.

[20] J. Dugan and J. Shubrook, "International Classification of Diseases, 10th Revision, Coding for Diabetes", Clinical diabetes: a publication of the American Diabetes Association, vol. 35, no. 4, pp. 232-238, 2017. doi:10.2337/cd16-0052.

# AI-Assisted UML Modeling for Serious Mental Illness Crisis Management

## Balancing Automation and Human Oversight: A Comparative Analysis of UML Diagramming Methods

Kerry Gilder
Applied Health Informatics Program
Fordham University
New York, USA
e-mail: kgilder@fordham.edu

John Chelsom
Applied Health Informatics Program
Fordham University
New York, USA
e-mail: jchelsom@fordham.edu

***Abstract*— This study assesses the effectiveness of Artificial Intelligence (AI) generated Unified Modeling Language (UML) diagrams in illustrating the treatment pathways for Serious Mental Illness (SMI) crises. AI-assisted tools, including ChatUML, ChatGPT, Claude, and DeepSeek, were assessed for accuracy, clarity, efficiency, and cost. A sample SMI use case was used to compare six AI-generated UML diagrams against a human-created benchmark. Results show that AI tools can improve diagram creation efficiency, with ChatUML using Claude 3.5 Sonnet and DeepSeek Reasoning with R1 accessed through Perplexity performing best. The limitations in the other AI-generated outputs demonstrate the need for human oversight to ensure precision in healthcare applications. The findings suggest that the use of generative AI in system planning and design would accelerate the development of pathways for healthcare providers managing SMI crises, with the potential to extend to other care settings.**

***Keywords-artificial intelligence; UML diagrams; serious mental illness; PlantUML; AI-assisted modeling.***

## I. Introduction

The management of SMI crises can be a challenge for healthcare providers and patients alike, requiring detailed organization and procedures to identify accurate care pathways [1]. To make this process clearer, the individual care teams can coordinate services using UML Use Case diagrams that serve as a blueprint for teams to communicate and track relevant data. These diagrams have been found to be supportive in understanding the interactions between healthcare providers and services.

UML has been proven to be an effective tool for visualizing software design. It has been a valuable method for graphical view of system relationships in software engineering. UML is published by the Object Management Group (OMG) currently as UML 2.5.1 as a standardized modeling language for software engineering and system design [2]. A review of 128 papers of UML applications by Koç et al. [3] concluded that UML was a useful aid for design, modeling and class diagrams that support the identification of development requirements and system scope for computer science and industry applications. The visual ability to quickly display a system's requirements has been shown to improve analysis and design by breaking down complex steps into manageable components [3]. A detailed guide to implementing complex medical information systems recommends using UML to simplify the modeling of component relationships and grouping activities within the stages of development lifecycle. UML diagrams show the relationship between users and the system, a key function in care pathway development where user roles can be complex in healthcare applications [4].

As AI tools became more available for developers, Cámara et al. [5] evaluated ChatGPT's ability to perform modeling tasks and act as a modeling assistant. They found that while there were enhancements provided to UML models, there were limitations including lack of consistency and syntactic and semantic issues. However, they reported that the correctness of the models produced when using ChatGPT with PlantUML was much higher than ChatGPT alone, where PlantUML models made fewer syntactic errors. They conclude with the encouragement of incorporating AI models into model-based software engineering for the betterment of the modeling profession. Their findings reinforce the experience in this study, where AI models with PlantUML produced reliable outcomes.

PlantUML is an open-source UML diagramming tool that allows a user to begin with a text-based language that provides simple model representation of complex systems that are often used in the healthcare sector. It is favored for its effectiveness for healthcare modeling tasks and proves to be reliable in being syntactically correct in healthcare applications. PlantUML's popularity is attributed to its support of all UML diagrams, including class, use case, activity, sequence, component, deployment, and object diagrams [5]. The iterative flexibility of PlantUML supports agile development methods that use incremental modeling without locking in costly upfront design decisions. A recent study on Large Language Model (LLM) generation of UML models reported that PlantUML was the preferred tool for AI assisted UML diagram creation due to its widespread use and representation in LLM training data [6].

In researching available tools for UML generation, ChatUML was identified as a publicly available tool that describes itself as an AI-powered diagram generator that allows users to create and edit PlantUML diagrams using natural language conversations. It was launched in 2020 and generates UML diagrams through conversational interfaces through the ChatUML website and provides initial credits at no cost, with fees for additional use [7]. Earlier versions provided only ChatGPT for AI assistance but currently it offers a selection of commercially available AI tools to create PlantUML code for diagram generation. In addition to OpenAI's ChatGPT 4o, Anthropic's Claude Sonnet 3.5 and DeepSeek's R3 models were chosen for this study as current available AI tools inside ChatUML. The study will also evaluate each AI tool outside of the ChatUML shell for comparative effectiveness. Performance metrics of UML output for the SMI application will include technical accuracy, diagram clarity, time efficiency, and user costs. A manual diagram will be created as a benchmark to compare with AI tool's added value. The SMI use case will be used to create a consistent prompt across all tools, with comparative metrics used to provide a recommended toolset for developing UML. The SMI use case was developed using the Medicaid Innovation Accelerator Program (IAP), which helps manage beneficiaries with complex care needs and assists state agencies with data and workflow management of beneficiaries with SMI [8].

Complex SMI crisis management requires clear care pathways that UML diagrams can assist in structuring. This study aims to assess AI UML modeling tools for their effectiveness in generating accurate, efficient, and usable diagrams. This research will identify reliable and cost-effective AI-assisted solutions by comparing six AI-generated UML use case diagrams against a manual benchmark. The goal is to recommend an accessible AI tool that supports both system developers and healthcare providers in improving care coordination, decision-making, and workflow efficiency within the SMI treatment framework.

As AI has rapidly evolved, its tendency to hallucinate and generate false output that misrepresents the intended prompt has increased. Specific barriers with ethical, technological, liability and regulatory, workforce, social, and patient safety concerns have been cited, with the conclusion that human intervention is required to address barriers before AI can be safely and successfully applied widely in healthcare settings [9]. A Human-In-The-Loop (HITL) approach to augmenting AI in healthcare applications has been described, with human guided expertise to ensure safe application of AI in healthcare to lessen the fears and concerns that can stifle adoption of enabling technology [10]. Maintaining a human in the loop for AI applications is both a control for AI errors and a method of enhancing AI effectiveness in healthcare applications.

The objective of this research is to measure and validate the speed and efficiency that AI tools can provide in supporting and enhancing the human generation of UML diagramming that balances automation with human oversight to ensure accuracy and compliance with healthcare standards.

This investigation is organized to include a review of related works and an exploration of current approaches in the field of healthcare modeling in Section II. It then outlines the methodology for a comparative analysis in Section III, presents the results in Section IV with observations in Section V, and concludes with an evaluation of the findings and opportunities for future improvement in Section VI.

## II. Related Works

The use of Unified Modeling Language (UML) in healthcare settings has a history of studies that show the value in visually demonstrating system processes and provider interactions [3]. Recently, the integration of AI into modeling tasks has significantly advanced. Cámara et al. [5] explored ChatGPT's performance with UML modeling and noted improved outcomes when combined with PlantUML, while noting minor syntax issues. Conrady and Cabot [6] displayed how Large Language Models (LLMs) can generate UML diagrams from visual prompts in their work. These studies suggest that while AI can enhance modeling efficiency, there are still accuracy and consistency limitations.

The current study leverages these findings and focuses on the application of AI-generated UML diagrams in behavioral health, particularly for Serious Mental Illness (SMI) crisis management. It also includes multiple LLMs and use environments (ChatUML vs. native interfaces), as well as a benchmarking process against a human-generated diagram.

AI-assisted UML generation can utilize several tools, such as ChatGPT, Claude, and DeepSeek, as explored here. These tools, along with others, vary in their accuracy and usability. Prior studies have explored UML generation in generic modeling contexts [5], [6], but their application to healthcare-specific use cases or high-stakes environments such as serious mental illness (SMI) crises is less common. In the behavioral health use case explored here, existing solutions lack precision and struggle to perform consistently across dynamic, multi-role scenarios. A notable limitation of these tools is their current inability to consistently meet regulatory requirements and standards, making them unreliable in real-world application.

This study evaluates how current AI tools perform when applied to a standardized behavioral health use case. The ability to use these tools in real time with human input is also considered. The goal is to determine whether these tools, as they develop, can reliably support SMI response planning.

## III. Methods

To meet the research objective of determining the practicality of using emerging AI tools to generate a UML use case diagram for the care of a client experiencing an SMI crisis, the approach was to employ data simulation, manipulation, and processing. IAP's list of select behavioral health procedures for adult Medicaid beneficiaries with SMI was extracted for the simulated data input. For this evaluation, the procedures were divided into two categories, provider type and procedure type.

This data set was used as prompts across six AI tool sets, including ChatGPT (inside and outside ChatUML), Claude Sonnet 3.5 (inside and outside ChatUML), and DeepSeek (inside ChatUML and Perplexity). ChatGPT was selected with previous research documenting its UML capability [5] as well as its commitment to ensure privacy where specified [11]. Anthropic's Claude was included in the evaluation for its reputation for research accuracy and code generation capabilities. Anthropic's Privacy Policy emphasizes alignment with applicable law to provide user safety, ensuring user rights to delete, correct, restrict or withdraw consent for use [12]. DeepSeek has recently documented efficiency and low development costs as well as strong coding proficiency [13]. DeepSeek was not accessed directly through the standalone Chinese hosted servers due to potential security concerns with Chinese privacy practices. To access native DeepSeek Reasoning with R1, Perplexity has provided hosting service on US and Canadian servers that provide a higher level of security. Perplexity's Terms of Use allow users to define Confidential Information and restrict its use and commit to complying with the EU-U.S. Data Privacy Framework and the UK Extension, certified by the U.S. Department of Commerce [14]. The comparison of DeepSeek V3 within ChatUML to DeepSeek Reasoning with R1 outside of ChatUML using Perplexity was chosen to evaluate DeepSeek outside of ChatUML as an AI tool option for this research. The diagrams that were generated outside ChatUML required an additional step of taking the PlantUML code to a PlantText editor. This step was accounted for in the time elapsed consideration during scoring.

Figure 1 illustrates the dual-path procedure used to generate the UML diagrams. The prompts are input into either ChatUML or a separate AI tool, with outputs rendered into diagrams via ChatUML directly or through the PlantText UML Editor. Testing was conducted on an LG Gram laptop with an Intel Evo i7 processor and 16GB RAM using a stable internet connection. Each test was repeated three times, and average response times were recorded. Subsequently, a human-generated UML diagram was created to provide a benchmark of each criteria.

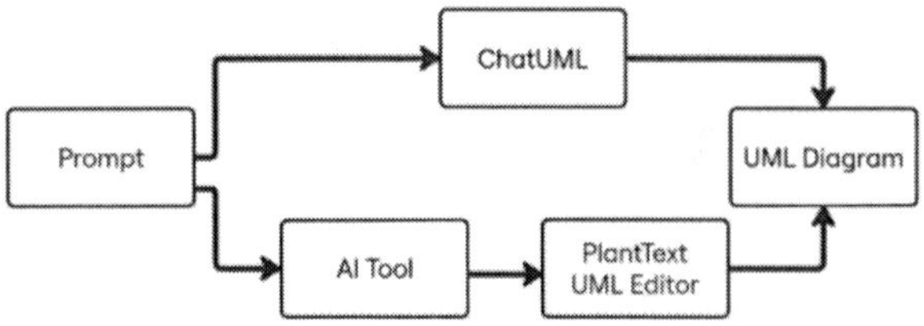


Figure 1: Creating UML diagram using AI tools.

To assess the performance of each tool, a manual review was conducted by a trained substance abuse counselor with knowledge of behavioral health workflows. Each AI-generated diagram was evaluated based on four key metrics, weighted according to their relative importance: Technical Accuracy (40%), Diagram Clarity (30%), Time Efficiency (20%), and User Cost (10%).

- Technical Accuracy – Measures the adherence of the UML diagram to the relationships provided and instructions explicitly defined in the prompt. 1 = fails to align with identified relationships or primary prompt instruction; 5 = accurately shows all provider relationships and workflows with no errors.
- Diagram Clarity – Ability to provide easily readable and usable diagrams. Clear diagrams enhance usability, enabling stakeholders to understand and use the models effectively. 1 = diagram is disorganized or unreadable, diagram fails to capture context; 5 = diagram is well-organized with logical flow, good labels and no distracting flows.
- Time Efficiency – Measures the total elapsed time for creating final UML output beginning with prompt initiation and includes iterations for correction and refinements. The quicker the tool generates accurate diagrams, the more efficiently it can support iterative workflows and real-time decision-making during development. 1 = over 60 seconds; 5 = Under 10 seconds.
- User Cost – Expense required for completing each output. This measure is a relative comparison of costs for each tool. 1 = most expensive; 5 = free.

Technical Accuracy was the highest weighted metric to assure proper use. Diagram Clarity was second to assure ease of use. Time Efficiency was third and included iterations, with human-generated results serving as a comparison. User Cost was the lowest weight, serving as a relative comparison of nominal expenses across tools. Each diagram was scored using a five-point scale per metric, and an overall weighted score was calculated using the formula:
Final Score=(0.4×Accuracy Score)+(0.3×Clarity Score)+(0.2×Time Efficiency Score)+(0.1×User Cost Score)

This methodology ensured a structured evaluation, allowing for a direct comparison of AI-generated UML diagrams against each other and the human-generated benchmark. The results provide comparisons of accuracy, clarity, efficiency, and cost-effectiveness across different AI models.

Prompt engineering was important for this study, using behavioral health experience and AI to craft prompts for SMI-related workflows. The prompt was designed and the procedures were classified in groups using knowledge of the Diagnostic and Statistical Manual of Mental Disorders, Fifth

Edition, Text Revision (DSM-5-TR) [15] and behavioral health treatment models. Additionally, successive iterations were applied to assess how different wording and structuring influenced AI outputs. Due to space limitations, only two packages of the diagram are illustrated as the practical example; however, the outcome is consistent with the use of the full prompt that included seven packages.

Original Prompt (seven packages): Create a vertical use case diagram with the information provided, please use plantuml Core Psychiatric Services 1. Pharmacologic Management * Clinician: Psychiatrist 2. Other Psychiatric Services/Procedures * Clinician: Psychiatrist 3. Diagnostic Interview Exam * Clinician: Psychiatrist, Psychologist 4. Crisis Intervention * Clinician: Psychiatrist, Crisis Counselor Therapeutic Services 1. Individual Therapy * Clinician: Therapist 2. Group Psychotherapy * Clinician: Therapist 3. Behavioral Health Day Treatment * Clinician: Therapist 4. Individual Psychotherapy * Clinician: Therapist 5. Family Psychotherapy * Clinician: Marriage & Family Therapist (MFT) Assessment & Diagnosis 1. Mental Health Assessment * Clinician: Psychologist, Therapist 2. Diagnostic Interview Exam * Clinician: Psychiatrist, Psychologist 3. Alcohol/Drug Assessment * Clinician: Addiction Counselor, Psychiatrist Substance Use Disorder Services 1. Alcohol/Drug Services Intensive Outpatient * Clinician: Addiction Counselor, Case Manager 2. Alcohol/Drug Abuse Services * Clinician: Addiction Counselor 3. Alcohol/Drug Case Management * Clinician: Case Manager Crisis & Community-Based Care 1. Crisis Intervention * Clinician: Crisis Counselor, Psychiatrist 2. Community-based Wrap-around Services * Clinician: Case Manager, Therapist Medical & Physical Health 1. New Patient Office Visit * Clinician: Physician 2. Admission History & Physical Exam * Clinician: Physician, Psychiatrist Behavioral & Preventive Services 1. Therapeutic Behavioral Services * Clinician: Behavioral Therapist 2. Behavioral Health Prevention Education * Clinician: Health Educator 3. Behavioral Health Prevention Information Dissemination * Clinician: Health Educator [8].

Truncated Prompt for the Practical Example (two packages): Create a vertical use case diagram with code for the information provided, be sure to include a heading. Please use PlantUML: Core Psychiatric Services 1. Pharmacologic Management * Clinician: Psychiatrist 2. Other Psychiatric Services/Procedures * Clinician: Psychiatrist 3. Crisis Intervention * Clinician: Psychiatrist, Crisis Counselor Assessment & Diagnosis 1. Mental Health Assessment * Clinician: Psychologist, Therapist 2. Diagnostic Interview Exam * Clinician: Psychiatrist, Psychologist 3. Alcohol/Drug Assessment * Clinician: Addiction Counselor, Psychiatrist.

## IV. RESULTS

Figures 2 – 8 present the UML use case diagram outcomes of the standardized truncated prompt for each of the six AI tool methods described, and the human created diagram. Table 1 presents the scoring for each method for the weighted criteria described.

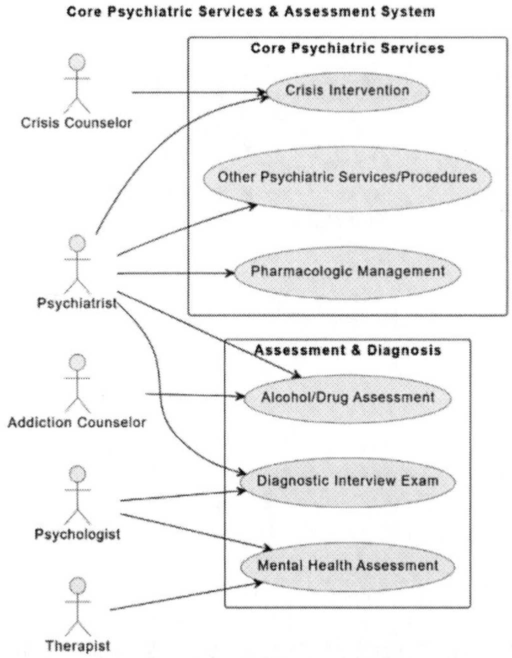


Figure 2: ChatUML, Claude 3.5 Sonnet.

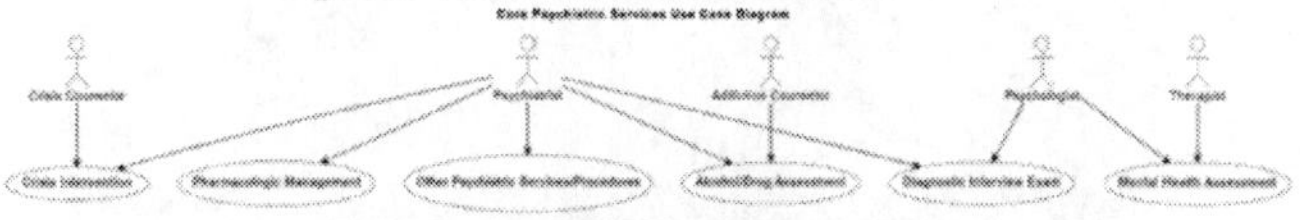

Figure 3: ChatUML, ChatGPT 4o.

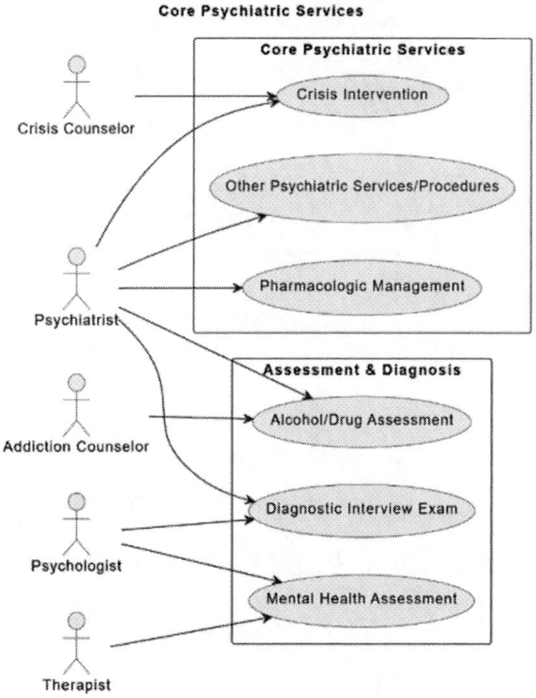


Figure 4: ChatUML, DeepSeek V3.

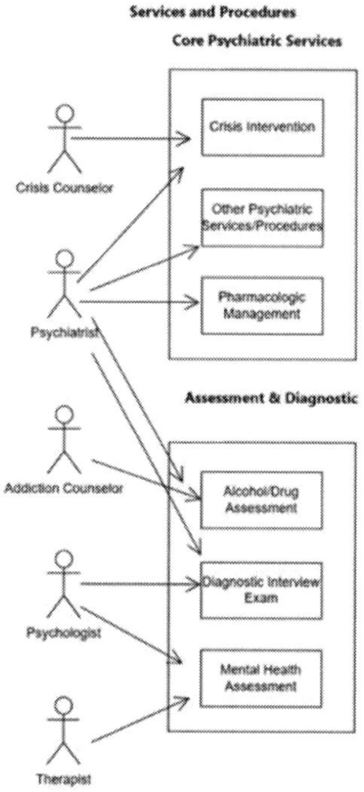


Figure 5: Human, experienced in behavioral health.

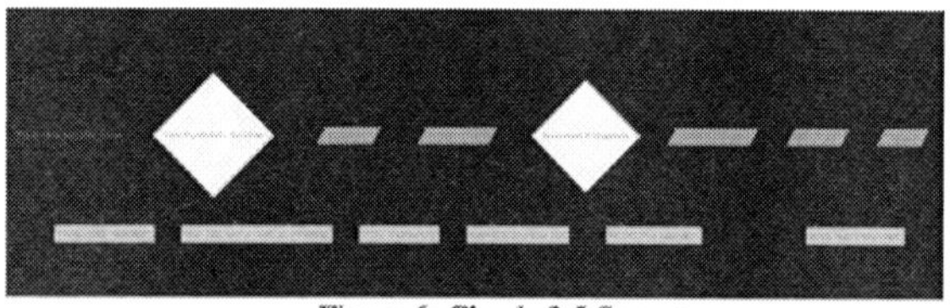

Figure 6: Claude 3.5 Sonnet.

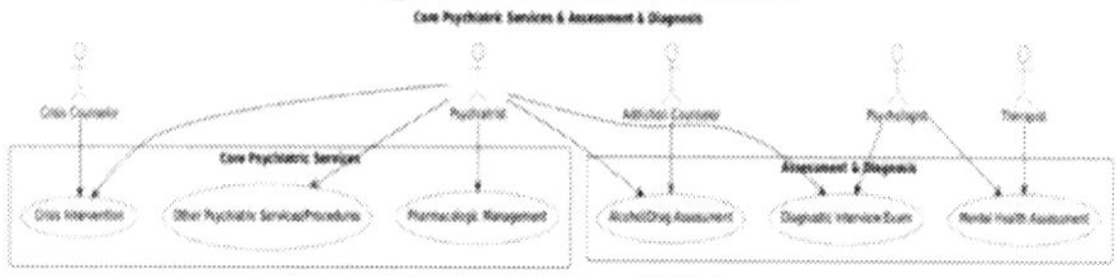

Figure 7: ChatGPT 4o.

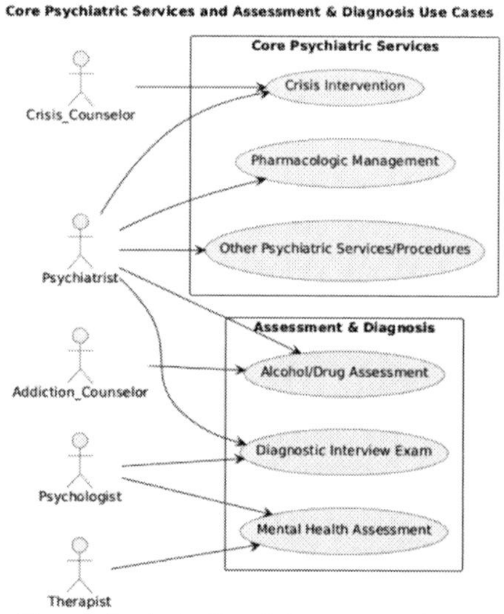


Figure 8: Perplexity, DeepSeek Reasoning with R1.

TABLE I. UML USE CASE DIAGRAM SCORECARD TABLE.

| AI Tool | Technical Accuracy | Diagram Clarity | Time Efficiency | User Cost | Weighted |
|---|---|---|---|---|---|
| ChatUML, Claude | 5 | 5 | 5 | 4 | 4.9 |
| ChatUML, ChatGPT | 2 | 2 | 5 | 4 | 2.8 |
| ChatUML, DeepSeek | 3 | 5 | 2 | 4 | 3.5 |
| Human | 5 | 5 | 1 | 5 | 4.2 |
| Claude | 1 | 2 | 4 | 5 | 2.3 |
| ChatGPT | 3 | 4 | 4 | 1 | 3.3 |
| DeepSeek, Perplexity | 5 | 5 | 4 | 5 | 4.8 |

The highest scoring method was Claude 3.5 Sonnet inside ChatUML, with a weighted score of 4.9, followed closely by DeepSeek Reasoning with R1 accessed by Perplexity with a score of 4.8. The lowest score was Claude Sonnet 3.5 used outside of ChatUML.

## V. DISCUSSION

The results revealed several issues within the Technical Accuracy criteria. ChatGPT, both within and outside ChatUML, as well as Claude Sonnet 3.5 outside ChatUML, did not follow the prompt. ChatGPT in both environments produced horizontal diagram outputs despite clear instructions to the contrary, as vertical was specified. Meanwhile, Claude Sonnet 3.5 in its native environment did not use PlantUML as directed and chose to default to Mermaid, which is Claude's built-in diagramming tool. Additionally, DeepSeek within ChatUML mislabeled the diagram by generating an incorrect heading. These deficiencies in Technical Accuracy are reflected in the scores. The Human, as expected, created an accurate diagram.

All AI tools were given the same prompt but their Diagram Clarity varied based on structure and presentation. ChatUML (Claude) and ChatUML (DeepSeek) produced the clearest diagrams, scoring 5/5. ChatUML (ChatGPT) scored the lowest (2/5) due to missing boxes and headings, making it difficult to follow. Native ChatGPT (4/5) was structured vertically, which affected readability. Native Claude (2/5) chose distracting colors and had small text, making it practically illegible. DeepSeek (Perplexity) and human-created diagrams (both 5/5) showing AI parity with human design. This highlights the importance of formatting, structure, and visual presentation in Diagram Clarity.

The Time Efficiency results show differences in processing times within AI tools, and the expected delays with the manual/human method. Both Claude Sonnet 3.5 and ChatGPT 4o performed efficiently within ChatUML, each completing the task in 10 seconds. However, DeepSeek took longer at 60 seconds, which may be due to inefficiencies between ChatUML and DeepSeek V3. The manual process, by comparison, was the most time-consuming at 30 minutes, demonstrating the advantage of automation. When tested natively, Claude Sonnet 3.5 maintained its 10-second speed, while ChatGPT 4o took slightly longer at 17 seconds. Interestingly, Perplexity DeepSeek performed much faster natively, completing the task in 15 seconds compared to its significantly longer runtime in ChatUML, perhaps indicating efficiencies provided by Perplexity. These findings show the potential trade-offs between different AI environments and suggest that while some tools perform optimally in certain settings, others may experience slowdowns due to integration constraints.

The cost analysis of UML implementations shows that ChatGPT Native is the most expensive option (rated 1), costing $20 per month. The middle tier (rated 4) consists of ChatUML implementations across platforms—Claude ChatUML, ChatGPT ChatUML, and DeepSeek ChatUML—offering a lower priced alternative. ChatUML operates on a credit-based system, with pricing starting at $2.99 for 20 credits, though the 250-credit package for $6.99 was used here due to the iterative approach and multiple tools sampled. It is worth noting that ChatUML follows a tiered system, where all three tools used in this analysis cost 3 credits per request. Previously, the credit cost for ChatGPT 4o and Claude 3.5 Sonnet was 5 credits per request but on December 7, 2024, it decreased to 3 credits. DeepSeek ChatUML, which was only integrated after December 2024, has always been priced at 3 credits per request. Finally, Human manual development, Claude Native, and DeepSeek/Perplexity Native are the most economical choices, as they are free (rated 5).

The outcome of the scoring against the weighted criteria indicates that Claude (ChatUML) is the recommended tool set for producing accurate, clear, efficient and cost-effective UML for an SMI application. Perplexity DeepSeek is also a reasonable tool set and is only slightly slower due to the extra step of copying PlantUML code into PlantText.

## VI. Conclusion

This study suggests that the application of AI in UML diagram generation holds significant potential to address the challenges of time efficiency and complexity management. This research started with the objective of evaluating artificial intelligence in the generation of a UML use case diagram that described a simulated care pathway for patients experiencing an SMI crisis that was managed by multiple healthcare providers. This objective was met through the practical evaluation of six AI tool environments. ChatUML and PlantUML are powerful tools for supporting the creation of a detailed diagram that met the research objective. Claude Sonnet 3.5 used inside of ChatUML provided the most accurate output and allowed for a more nuanced diagram of the interactions within the SMI crisis management system, demonstrating the potential of AI to significantly reduce the time and effort required in the system design process. The generated UML diagram (Figure 2) was a successful step in meeting the research goal. The investigation showed that AI can accelerate the development cycle and assist in managing complex system design. However, it is not yet ready to be hailed as the complete substitute for an informed expert. The AI-generated diagram required iterations and oversight to ensure that the details of SMI crisis management were accurately captured. This reinforces the current state of AI as a complementary tool rather than a replacement for human expertise. There are opportunities for further work that can expand on these research findings and add to its usefulness. As AI tools progress, it is anticipated that providing more domain-specific knowledge to AI Large Language Models (LLMs) like OpenAI, Claude and DeepSeek will enhance their ability to generate more scenario specific UML diagrams. AI itself will likely then be able to leverage the domain knowledge to automatically update diagrams based on real-time data, which could provide a dynamic tool for system management. Additionally, providing feedback of user's interactions with AI-generated diagrams could add to their effectiveness as communication tools for healthcare stakeholders, maintaining the HITL principle [10]. In reflection, the research investigation has confirmed that AI has current usefulness and significant promise for system design in the complex healthcare sector. There was human learning in the exercise as well. The iterations required by this research identified better ways to phrase a problem for the AI to solve. Healthcare providers as novice users will quickly become more agile in using ChatUML with Claude Sonnet 3.5 or DeepSeek Reasoning R1 through Perplexity, and their second and third applications will produce more useful diagrams. While the AI did not fully replace the need for human expertise, it served as an intelligent assistant by streamlining the design process. The results present a compelling case for the integration of AI in technical documentation practices as a viable way forward while maintaining control with a human oversight. As AI continues to evolve, its potential to transform system design and management practices is evident, validating continued exploration and development for healthcare design.

The combination of AI-powered methods produced a usable outcome as the final product. It demonstrates the efficiency provided by significantly reducing the time required to produce complex UML diagrams. The relative speed and efficiency of using AI tools is a meaningful contrast to the traditional manual diagramming and programming methods. This research identifies available public tools that provide advancement in the system design and documentation of UML diagrams for healthcare use cases. This application confirms the efficacy of AI in generating critical technical artifacts and demonstrates the possibilities for more rapid adaptation of a new approach for system visualization in the context of healthcare management.

## ACKNOWLEDGMENT

We thank Fordham University and the Applied Health Informatics program for their support. Kerry Gilder is especially grateful to her teachers, classmates, and her family, particularly her father, whose openness and willingness to listen have been invaluable in shaping her ideas and education.


## REFERENCES


[1] F. Röhricht et al., "Implementation of a novel primary care pathway for patients with severe and enduring mental illness," BJPsych Bulletin, vol. 41, no. 6, pp. 314-319, Dec. 2017, doi: 10.1192/pb.bp.116.055830.

[2] [Object Management Group (OMG), Unified Modeling Language (OMG UML), Version 2.5.1, 2017. [Online]. Available: https://www.omg.org/spec/UML/2.5.1/PDF, retrieved: January, 2025.

[3] H. Koç, A. M. Erdoğan, Y. Barjakly, and S. Peker, "UML diagrams in software engineering research: A systematic literature review," Proceedings, vol. 74, no. 1, p. 13, 2021. [Online]. Available: https://doi.org/10.3390/proceedings2021074013, retrieved: January, 2025.

[4] A.-M. Ştefan, N.-R. Rusu, E. Ovreiu, and M. Ciuc, "Empowering healthcare: A comprehensive guide to implementing a robust medical information system—Components, benefits, objectives, evaluation criteria, and seamless deployment strategies," Appl. Syst. Innov., vol. 7, no. 3, p. 51, Jun. 2024. [Online]. Available: https://doi.org/10.3390/asi7030051, retrieved: January, 2025.

[5] J. Cámara, J. Troya, L. Burgueño and A. Vallecillo, "On the assessment of generative AI in modeling tasks: an experience report with ChatGPT and UML," Softw. Syst. Model., vol. 22, pp. 781–793, 2023. [Online]. Available: https://doi.org/10.1007/s10270-023-01105-5, retrieved: January, 2025.

[6] A. Conrady and J. Cabot, "From image to UML: First results of image-based UML diagram generation using LLMs," arXiv:2404.11376 [cs.SE], 2024. [Online]. Available: https://doi.org/10.48550/arXiv.2404.11376, retrieved: January, 2025.

[7] ChatUML, "ChatUML," [Online]. Available: https://chatuml.com/, retrieved: January, 2025.

[8] Centers for Medicare & Medicaid Services, "Using data analytics to better understand Medicaid populations with serious mental illness," Medicaid Innovation Accelerator Program, n.d. [Online]. Available: https://www.medicaid.gov/state-resource-center/innovation-accelerator-program/iap-downloads/program-areas/data-analytics-smi-tech-resource.pdf, retrieved: January, 2025.

[9] M. I. Ahmed, B. Spooner, J. Isherwood, M. Lane, E. Orrock, and A. Dennison, "A systematic review of the barriers to the implementation of artificial intelligence in healthcare," Cureus, vol. 15, no. 10, p. e46454, 2023. [Online]. Available: https://doi.org/10.7759/cureus.46454, retrieved: January, 2025.

[10] E. Sezgin, "Artificial intelligence in healthcare: Complementing, not replacing, doctors and healthcare providers," Digital Health, vol. 9, p. 20552076231186520, 2023. [Online]. Available: https://doi.org/10.1177/20552076231186520 retrieved: January, 2025.

[11] OpenAI, "Usage policies," 2024. [Online]. Available: https://openai.com/policies/usage-policies/, retrieved: January, 2025

[12] Anthropic, "Privacy policy," 2024. [Online]. Available: https://www.anthropic.com/legal/privacy, retrieved: January, 2025.

[13] DeepSeek AI, "DeepSeek-V3 technical report," arXiv, Dec. 26, 2024. [Online]. Available: https://arxiv.org/abs/2412.19437, retrieved: January, 2025.

[14] Perplexity AI, "Enterprise terms of service," 2025. [Online]. Available: https://www.perplexity.ai/hub/legal/enterprise-terms-of-service, retrieved: January, 2025.

[15] American Psychiatric Association, Diagnostic and Statistical Manual of Mental Disorders, 5th ed., text rev., 2022. [Online]. Available: https://doi.org/10.1176/appi.books.9780890425787, retrieved: January, 2025.

# Enhancing Fall Prediction in Older Adults: A Data-Driven Approach to Key Parameter Selection

Amadou M. Djiogomaye Ndiaye[a, e], Michel Harel[a,b,g], Laurent Billonnet[f], Achille Tchalla[a, c,d, e]
[a]Laboratoire Vie-Santé, UR 24 134, Faculté de Médecine, 87025 Limoges, France,
[b]INSPÉ de Limoges, Université de Limoges, 87036 Limoges Cedex, France; UMR5219 UPS
[c] Service de Médecine Gériatrique, CHU de Limoges Dupuytren, Limoges, France,
[d]Pôle Hospitalo-Universitaire de Gérontologie clinique, Service de Médecine Gériatrique, Unité de Prévention de Suivi et d'Analyse du Vieillissement, CHU de Limoges Dupuytren, Limoges, France
[e]Chaire d'excellence chez Fondation Partenariale de l'Université de Limoges I Institut Omega Health, Laboratoire VieSanté, UR 24134 Vieillissement Fragilité Prévention e-Santé
[f]XLIM-SRI - Systèmes et Réseaux Intelligents (444315) - XLIM - 123, avenue Albert Thomas - 87060 LIMOGES CEDEX - France
[g]Institut de Mathématiques de Toulouse, UMR CNRS 5 219, Université Paul Sabatier, Toulouse 31 062, France,
e-mail: amadou_maguette_djio.ndiaye@unilim.fr,
michel.harel@unilim.fr,
laurent.billonnet@unilim.fr,
achille.tchalla@chu-limoges.fr

*Abstract*— **Falls significantly contribute to frailty and functional decline in elderly individuals. The Risk Of Falling (ROF) is linked to three dimensions: physical/organic, socio-environmental, and thymic/cognitive. Therefore, fall prevention protects older individuals from multiple comorbidities. The reliability of predictive studies depends on the quality and consistency of data collection. In most studies, data for model construction were collected from hospitals, research laboratories, or participants' homes. Recent fall prediction models increasingly rely on machine learning, deep learning, and computer vision. Predictive models assist specialists in decision-making. Using home-collected data, our objective is to develop an optimized predictive model with minimal features. In this paper, we aim to identify the optimal model for predicting falls using this strategy with the objective of building a robust dataset as the entry of an Artificial Intelligence (AI) process.**

***Keywords - fall; older population; data; prevention; AI.***

## I. Introduction

According to the World Health Organization (WHO), older individuals are those aged ≥ 60 years. The proportion of older individuals worldwide is expected to nearly double between 2015 and 2050, increasing from 12% to 22% [1]. The National Institute of Statistics and Economic Studies (INSEE) estimates that one in three individuals in France will be aged ≥ 60 years by 2060, compared to one in four individuals in 2021 [2]. Aging leads to a gradual decline in functional capacity, increasing the ROF [3]. Early identification of ROF facilitates the administration of personalized interventions for individuals [4].

Most recent studies predict falls using sensors or Electronic Health Records (EHRs). With data collected directly from elderly individuals' homes, our objective is to develop an effective predictive model using the fewest possible features. Our initial models, built with a comprehensive approach, have demonstrated satisfactory performance. In this work, we briefly discuss fall prediction strategies, the dimensions of ROF, and relevant studies in the field. In Section II, we define falls using a holistic approach. In Section III, we discuss the findings from previous studies as well as the best predictive model we developed. Finally, in Section IV, we present a concise and synthetic conclusion.

## II. Risk of Falling

Falls can be caused by various factors. A fall occurs when a person involuntarily lands on the ground or at a lower level than their starting position [5]. It can occur multiple times a year. Geriatrics who experience at least two falls within 12 months are classified as "fallers" [6].

A holistic fall prediction approach considers three key dimensions:

- The physical/organic dimension gathers data related to an individual's medical history and current symptoms, diagnosis of underlying health issues, and treatment effectiveness.
- The thymic/cognitive dimension refers to an individual's mental, emotional, and cognitive states.
- The socio-environmental dimension refers to age, gender, family and social support, housing conditions, home configuration, the presence of slippery rugs, stairs without railings, uneven surfaces, and inadequate lighting.

Evaluating the ROF involves at least a gait and balance assessment of the physical/organic dimension and the age and gender of the socio-environmental dimension. Data involving the thymic/cognitive dimension allows for a comprehensive review of the potential causes of a fall. The term "dimension" refers to the types of factors that contribute to the ROF and their evaluation.

Hospitalized patients often receive incomplete health assessments across all dimensions. Our home-collected data encompasses features from all three dimensions.

## III. PREDICTIVE FACTORS FOR FALLS IN THE OLDER POPULATION

Not every feature within the three ROF dimensions is a predictive factor for falls. The effectiveness of a predictive factor depends on its statistical significance, correlation with fall occurrences, and its interaction with other variables across the physical/organic, socio-environmental, and cognitive dimensions. In some studies, the identified predictive variables did not encompass all three dimensions of ROF. Kawazoe et al. [7], Ikeda et al. [8], and Cella et al. [9] demonstrated that age category related to socio-environmental was a predictor of falls, suggesting a strong association between age and falls. Bath et al. found that the predictive variables related to the socio-environmental dimension are diverse and varied, contributing to effective prevention [10]. In fact, a higher number of variables related to gait and balance is associated with a more robust predictive model for falls.

In the literature review conducted by Rubenstein, only cognitive impairment was identified as a predictive variable related to the thymic/cognitive [11]. Conversely, Ikeda et al. [8], Kawazoe et al. [7], and Bath et al. [10] identified at least two predictive variables involving the thymic/cognitive dimension, providing a better understanding of the ROF associated with the thymic/cognitive dimension and facilitating preventive measures. In those features, we can find fear of falling, depressive symptoms, self-rated health, impaired consciousness, and dementia at admission. Recent studies by Ikeda et al. [8] and Kawazoe et al. [7] achieved Area Under the receiver operating characteristic Curve (AUC) scores of 88% and 85%, respectively, using comprehensive approaches. Ikeda et al. employed a Random Forest-based Boruta algorithm for feature selection, while Kawazoe et al. used a combination of Bidirectional Encoders and Bidirectional Long Short-Term Memory (BiLSTM) networks to process sequential data. These AUC scores indicate strong model performance, reflecting high discriminative ability in classification tasks [12].

The reliability of predictive studies depends on the quality and consistency of data collection. Unlike hospitals, where data is gathered only when patients seek care, home-based data collection requires practitioners to schedule visits at fixed intervals. Our study followed 1,648 community-dwelling older adults ($\geqslant$60 years) between September 2011 and September 2023. Participants were assessed at home by the Unit for Prevention, Monitoring and Analysis of Ageing (UPSAV – *Unité de Prévention, de Suivi et d'Analyse du Vieillissement*) at Limoges University Hospital, Limoges, France. Each patient underwent an initial visit followed by a second visit six months later. After the second visit, annual follow-ups were conducted for up to six years, as long as the patient remained in their home. Data collection included cardiovascular risk factors, fall occurrences, socio-environmental characteristics, and a comprehensive geriatric assessment summary score. To ensure coherence, our predictive model accounts for temporality by predicting falls every six months. Out of thirty input features, eleven were identified as the most relevant fall predictors using multinomial logistic regression: Gender, Hypertension, Obesity, Activities of Daily Living (ADL), Mini-Mental State Examination (MMSE), Short Physical Performance Battery (SPPB), Pathological Geriatric Depression Scale (GDS), Instrumental Activities of Daily Living (IADL), Leisure activity, Pathological Single-Leg Balance (SLB), and history of falling in the past year.

Of the 1,648 patients assessed at the first visit, 954 remained for the second visit. The AI model was subsequently built using data from these 954 patients. The most relevant fall predictors were used to develop several predictive models, including Logistic Regression, Support Vector Machine (SVM), Random Forest, and eXtreme Gradient Boosting (XGBoost). The area under the receiver operating characteristic curve (AUC) was used to evaluate the predictive performance of these models. Among them, Random Forest and XGBoost achieved the highest performance, with AUCs of 77 and 76, on the test set. XGBoost achieved a Brier score of 0.19, while that of Random Forest was 0.2.

The Brier score measures the accuracy of probabilistic predictions; it is calculated as the mean squared difference between predicted probabilities and actual outcomes. Lower values indicate better calibration [13].

Due to its low Brier score and suitability for interpretation, the XGBoost model was selected for SHapley Additive exPlanations (SHAP) analysis to interpret its predictions at the individual level. SHAP analysis revealed that fall history, balance performance, cognitive status, and functional ability were the most influential predictors.

SHAP is an explainable AI method that provides insights into the contribution of each feature both globally (across the entire dataset) and locally (for individual predictions) [14].

As with most AI models, ours can be continuously refined with additional data over time. In our case, improving the model also provides an opportunity to collect data from patients' homes while offering them personalized fall prevention advice. During the intervals between practitioner visits, necessary adjustments to home configurations can also be made if needed.

The limitations of this study include the use of data that did not account for medications taken by participants or treatments for specific comorbidities, which may influence fall risk. Additionally, the study population was limited to older adults residing in France, potentially affecting the generalizability of the findings to other geographic or cultural contexts. Future research should aim to include more diverse populations to enhance the external validity and applicability of the results.

## IV. Conclusion

This study contributes to advancing fall prevention by leveraging a 12-year dataset collected in home settings to develop an AI-based predictive model. Our approach integrates the three dimensions of ROF, optimizing model performance while reducing the number of required input features.

By applying explainable AI techniques, we identified the contribution of each feature to fall risk, thereby supporting the development of more targeted and effective intervention strategies. These findings may help enhance the quality of elderly care by informing personalized prevention efforts and guiding future research in geriatric risk assessment.

## References


[1] WHO, "Ageing and health" [Online] Available: https://www.who.int/fr/news-room/fact-sheets/detail/ageing-and-health, Accessed: Mar. 28, 2025.

[2] Insee, "Life Expectancy at Various Ages | Insee." [Online] Available: https://www.insee.fr/fr/statistiques/2416631#graphique-figure1, Accessed: Mar. 27, 2025.

[3] X. Thierry, "Accidents and physical assaults among the elderly: less frequent than among the young, but more severe," *Population & Sociétés*, vol. 468, no. 6, pp. 1–4, 2010, doi: 10.3917/popsoc.468.0001.

[4] US Preventive Services Task Force, "Interventions to Prevent Falls in Community-Dwelling Older Adults: US Preventive Services Task Force Recommendation Statement," *JAMA*, vol. 319, no. 16, pp. 1696–1704, Apr. 2018, doi: 10.1001/jama.2018.3097.

[5] S. L. Wolf, H. X. Barnhart, N. G. Kutner, E. McNeely, C. Coogler, and T. Xu, "Reducing frailty and falls in older persons: an investigation of Tai Chi and computerized balance training. Atlanta FICSIT Group. Frailty and Injuries: Cooperative Studies of Intervention Techniques," Journal of the American Geriatrics Society, vol. 44, no. 5, May 1996, doi: 10.1111/j.1532-5415.1996.tb01432.x.

[6] M. E. Tinetti and M. Speechley, "Prevention of Falls among the Elderly," http://dx.doi.org/10.1056/NEJM198904203201606. [Online]. Available: https://www.nejm.org/doi/pdf/10.1056/NEJM198904203201606, Accessed: Apr. 24, 2025.

[7] Y. Kawazoe, K. Shimamoto, D. Shibata, E. Shinohara, H. Kawaguchi, and T. Yamamoto, "Impact of a Clinical Text-Based Fall Prediction Model on Preventing Extended Hospital Stays for Elderly Inpatients: Model Development and Performance Evaluation," JMIR Med Inform, vol. 10, no. 7, p. e37913, Jul. 2022, doi: 10.2196/37913.

[8] T. Ikeda et al., "An Interpretable Machine Learning Approach to Predict Fall Risk Among Community-Dwelling Older Adults: a Three-Year Longitudinal Study," J Gen Intern Med, vol. 37, no. 11, pp. 2727–2735, Aug. 2022, doi: 10.1007/s11606-022-07394-8.

[9] A. Cella et al., "Development and validation of a robotic multifactorial fall-risk predictive model: A one-year prospective study in community-dwelling older adults," PLOS ONE, vol. 15, no. 6, p. e0234904, June 2020, doi: 10.1371/journal.pone.0234904.

[10] P. A. Bath, N. Pendleton, K. Morgan, J. E. Clague, M. A. Horan, and S. B. Lucas, "New approach to risk determination: development of risk profile fornew falls among community-dwelling older people by use of a GeneticAlgorithm Neural Network (GANN)," The Journals of Gerontology: Series A, vol. 55, no. 1, pp. M17–M21, Jan. 2000, doi: 10.1093/gerona/55.1.M17.

[11] L. Z. Rubenstein, "Falls in older people: epidemiology, risk factors and strategies for prevention," Age and Ageing, vol. 35, no. suppl_2, pp. ii37–ii41, Sep. 2006, doi: 10.1093/ageing/afl084.

[12] A. P. Bradley, "The use of the area under the ROC curve in the evaluation of machine learning algorithms," Pattern Recognition, vol. 30, no. 7, pp. 1145–1159, Jul. 1997, doi: 10.1016/S0031-3203(96)00142-2.

[13] L. Hoessly, "On misconceptions about the Brier score in binary prediction models," arXiv: arXiv:2504.04906. doi: 10.48550/arXiv.2504.04906, Apr. 23, 2025.

[14] S. M. Lundberg and S.-I. Lee, "A Unified Approach to Interpreting Model Predictions," in Advances in Neural Information Processing Systems, Curran Associates, Inc., 2017. [Online]. Available: https://proceedings.neurips.cc/paper/2017/hash/8a20a8621978632d76c43dfd28b67767-Abstract.html, Accessed: Apr. 20, 2025.

# Factors Influencing User Satisfaction in Electronic Health Record Systems

## A Regression Analysis and its Relevance to the Valkyrie Project

[a] Ove Lintvedt[1,2], [b] Espen S. Nordheim[1], [c] Terje Geir Solvoll[1,2], [d] Rune Pedersen[1,2]

[1]Norwegian Centre for E-health Research,
University Hospital of North Norway,
Tromsø, Norway
[2]Norway Faculty of Nursing and Health Sciences
Nord University
Bodø, Norway

e-mail: [a]ove.lintvedt@ehealthresearch.no, [b]espen.solbakken.nordheim@ehealthresearch.no,
[c]terje.solvoll@ehealthresearch.no, [d]rune.pedersen@ehealthresearch.no

***Abstract*—User satisfaction with Electronic Health Record (EHR) systems is critical for ensuring efficiency, usability, and quality of healthcare services. This study examines how training satisfaction, functional satisfaction, and general satisfaction influence overall satisfaction with an EHR system. Using regression analysis, we identify key predictors of user satisfaction and provide recommendations for system improvement. Results indicate that efficiency, usability, and workflow optimization are the most significant determinants of overall satisfaction, with training having a secondary but notable effect. Furthermore, the findings are linked to the Valkyrie project, a distributed service-oriented architecture designed to enhance healthcare coordination and data sharing. We explore how insights from this study can inform the development of the Virtual Health Record (VHR) in Valkyrie, highlighting the need for data accessibility and user-centred design.**

***Keywords-electronic health records; usability; workflow optimization; healthcare efficiency; Valkyrie project; virtual health record; interoperability; healthcare coordination.***

## I. Introduction

EHR systems are widely implemented to improve healthcare documentation [1], patient safety [2], and workflow efficiency [3]. However, user satisfaction remains a challenge, as healthcare professionals often encounter usability issues [4], workflow inefficiencies [5], and training limitations [6]. This study explores the relationship between training, functional, and generic satisfaction with EHR systems and their impact on overall user satisfaction.

The relevance of this study extends to the Valkyrie project [7], which aims to enhance healthcare coordination through a VHR that integrates data across multiple healthcare providers. The Norwegian government has promoted the development of EHRs that improve information flow [8]. Healthcare professionals frequently encounter challenges in navigating and interpreting the extensive volumes of patient data within EHR systems [9]. These challenges are often related to data overload and fragmentation, complicating the diagnostic process. Additionally, the design of many EHRs has been criticized for not adequately considering the complex cognitive and collaborative aspects of healthcare delivery. This oversight can lead to cumbersome systems for clinicians, further hindering efficient data navigation and interpretation [10].

A VHR is an extension of EHR systems, focusing on interoperability, real-time data access, and comprehensive patient insights. It represents a shift from institution-centric record-keeping to a patient-centred, data-driven approach, enabling improved clinical decisions and patient outcomes. The Valkyrie project and VHR have the same aims as the Patient-Centered Data Home (PCHD) [11], which aims to improve healthcare interoperability and patient data accessibility, but they differ in their architecture, implementation, and scope. The PCDH is a Health Information Exchange (HIE) model designed to facilitate the secure and automated exchange of patient health records across regional and national HIE networks. When patients receive care outside their primary healthcare system, the PCDH framework ensures that their home HIE automatically provides relevant medical history to the treating provider, enhancing care continuity and reducing duplication of tests and treatments. VHR and PCHD address healthcare interoperability challenges and improve patient-centred care by ensuring critical patient information is accessible across healthcare institutions. However, VHR is a broader system, dynamically aggregating real-time data from multiple sources, while PCHD is a structured, federated exchange model designed explicitly for secure HIE-based data sharing when patients move across different care settings. By leveraging VHR's real-time data accessibility and PCHD's structured HIE-based data exchange, healthcare systems can create a more interconnected, efficient, and patient-centred health information ecosystem.

By examining the key factors influencing EHR user satisfaction, this research provides insights regarding the improvement of user experiences in distributed electronic health record systems like Valkyrie.

This paper is structured as follows: Section II presents the methods, including setting, study design and data collection, variables and measures, and analysis techniques. Section III presents the results. Section IV discusses the findings. Finally, Section V is the paper's conclusion, with recommendations for future research.

## II. METHOD

### A. Related Work

Numerous studies have explored user satisfaction in EHR systems. Lintvedt et al. [12] investigated satisfaction post-deployment in Northern Norway, emphasizing usability and clinical documentation. Other research has focused on the influence of training and technical support on satisfaction [6] or explored how EHR systems affect patient safety and workflow efficiency [2][5].

Despite this, few studies have simultaneously analysed training, functionality, and general satisfaction as separate constructs feeding into overall satisfaction. Moreover, no prior work has used this regression framework to connect findings directly with a future system architecture like Valkyrie. Our contribution thus lies in identifying practical satisfaction predictors and applying them to ongoing system design.

### B. Setting

Norway's hospital sector is governed by four Regional Health Authorities (RHA), which oversee healthcare services in the South-East, West, Central, and North regions. In 2021, all RHAs were transitioning to a new EHR system. However, the Northern Norway Regional Health Authority (Helse Nord RHA) was the first and only region to fully implement the new EHR system, transitioning from DIPS Classic to DIPS Arena across its hospitals.

This study focuses on the hospitals within Northern Norway, specifically the University Hospital of North Norway (UNN), Nordland Hospital (NLSH), and Finnmark Hospital (FSH). This region was selected based on its unique status as the only RHA that had fully implemented the new EHR system in 2021 and that prior user satisfaction research was carried out in this RHA. No additional selection criteria were applied. However, the findings should be interpreted considering the specific characteristics and operational context of hospitals in Northern Norway.

### C. Study Design and Data Collection

This study employs a cross-sectional survey-based approach. The hospitals wanted to send the survey to all employees, with no possibility to find who many respondents that got the invitation. Of the 629 total respondents, only 407 indicated active use of the EHR system and completed the satisfaction sections. Responses from non-EHR users were excluded from the satisfaction and regression analyses, ensuring that all reported findings reflect the experiences of active EHR users. Data were collected from the healthcare professionals using a standardized questionnaire measuring satisfaction with different aspects of the EHR system. The survey used Likert scale responses (1 = total disagree to 5 = total agree). The participants were recruited through emails, with each hospital responsible for extending the invitation to all their employees. This method was considered as the most suitable solution as it used existing administrative structures. The hospitals did not want to send any reminders.

### D. Variables and Measures

The study analyses three primary categories influencing Overall Satisfaction (OS):

- Training Satisfaction (TS): Satisfaction with training (TS1), Perceived adequacy of training time (TS2), and Generic assessment of training quality (TS3).
- Functional Satisfaction (FS): Workflow and efficiency (FS1), Customization (FS2), Documentation and record-keeping (FS3), and Communication and patient interaction (FS4).
- Generic Satisfaction (GS): EHR efficiency for patient work (GS1), Clinical Quality Support (GS2), Automated Information Integration (GS3), Ease of use (GS4), and Support for clinical documentation (GS5), Clinical Documentation Support (GS6), Care Coordination Support (GS7).

In addition, data related to respondents' profession (physicians, nurses, and other professions), affiliation (three hospitals), gender, age, and clinical experience were collected.

### E. Statistical Analysis

Regression analyses were conducted to determine the relative contributions of TS, FS, and GS to OS. Stepwise regression was used to identify the most significant predictors. In addition, a regression was conducted to find a model for Overall Satisfaction based on the significant variables from the individual analyses. The statistical analysis was performed using Statistical Package for the Social Sciences (SPSS) 29.02 (IBM Corp., Armond, NY), with statistical significance set at $p < 0.05$.

There were missing data for the FS variables. For other variables, there were no missing values, as responses were mandatory. Missing values were addressed by applying the Missingness Completely At Random (MCAR) assumption by Little [13] and confirmed it ($\chi^2$=48.27, DF=47, p=.42). Expectation-Maximization (EM) analysis that estimates the means, covariances, and correlations was used to input missing values. The missing values were moderate, n=254 (15.60%), as some questionnaire items depended on the profession.

The statistical methodology aligns with prior studies on EHR user satisfaction [14][15], which employed regression analysis to identify key factors influencing user satisfaction in recently deployed EHR systems in Northern Norway. The use of regression to reduce satisfaction dimensions and assess overall usability effectiveness parallels our approach, reinforcing the robustness of our findings.

### F. Ethics

The study was presented and approved by the data-protection officer at the University Hospital of North Norway.

### G. Limitations

This study is subject to several limitations. Firstly, the voluntary nature of survey participation, the unknown number of invitees, and the absence of reminders may have led to self-selection bias, potentially attracting participants with strong opinions. Secondly, as the survey was distributed through hospital email systems without individual tracking, it is not feasible to verify the exact delivery rate or follow up on non-responses. Lastly, the study concentrates on hospitals within a single regional healthcare area, which limits the generalizability of the findings to other healthcare contexts.

## III. RESULTS

### A. Descriptives

The number of participants completing the primary survey was n=629, and participants who completed the satisfaction survey and were EHR users were n=407 (61.71%). Of this group, 63.30% were female. The average years of clinical experience was 18.03 years (sd=11.03), and the mean age of EHR users was 46.32 years (sd=11.42). In terms of professional roles among EHR users, physicians constituted n=101 (24.80%), nurses n=150 (36.90%), and other clinicians n=156 (38.30%). Age distribution within the EHR users was as follows: 6.60% (n=27) were between 18-29 years, 24.10% (n=98) were between 30-39 years, 25.6% (n=104) were between 40-49 years, 29.20% (n=119) were between 50-59 years, and 14.50% (n=59) were 60 years or older. The distribution among hospitals and professions is in Table I. The statistics for the satisfaction constructs is in Table II. A follow-up regression analysis was performed including covariates such as profession, gender, age, and clinical experience. Among these, clinical experience showed a weak positive correlation with overall satisfaction ($\beta = .12$, $p = .048$), while profession and age were not statistically significant. These findings suggest user satisfaction is primarily shaped by EHR functionality rather than demographic differences.

### B. Training Satisfaction and Overall Satisfaction

A linear regression was used to test if TS significantly predicted OS. The overall regression was statistically significant ($R^2 = .257$, $F(3, 403) = 46.378$, $p = <.001$).

Regression analysis of TS variables revealed that satisfaction with training quality (TS1) had the strongest impact on OS, see Table III. The $R^2$ value indicates a moderate explanatory power. The overall explained variance was modest, indicating that training alone does not sufficiently predict OS.

TABLE I. HEALTH REGION AND CLINICAL PROFESSIONS

| Health Region | Clinical profession | | | |
|---|---|---|---|---|
| | *Physicians, n* | *Nurses, n* | *Other, n* | *Total, n (%)* |
| FSH | 6 | 8 | 7 | 21 (5.26%) |
| NLSH | 37 | 77 | 75 | 189 (46.4%) |
| UNN | 58 | 65 | 74 | 197 (48.4%) |
| Total, n (%) | 101 (24.8%) | 150 (36.9%) | 156 (38.3%) | 407 (100%) |

TABLE II. DESCRIPTIVE STATISTICS FOR SATISFACTION CONSTRUCTS

| Satisfaction type | Descriptive statistics | |
|---|---|---|
| | *Mean* | *SD* |
| Training Satisfaction (TS) | 3.36 | 1.12 |
| Functional Satisfaction (FS) | 3.34 | .69 |
| Generic Satisfaction (GS) | 3.65 | .79 |
| Overall Satisfaction (OS) | 3.53 | .99 |

### C. Functional Satisfaction and Overall Satisfaction

A linear regression was used to test if FS significantly predicted OS. The overall regression was statistically significant ($R^2 = .508$, $F(4, 402) = 103.687$, $p = <.001$). The significant predictors are listed in Table III. The strongest predictors were FS1, FS2 and FS3. The FS4 standardized coefficient ($\beta = -.151$) indicates a weak negative effect, meaning Communication and patient interaction have a small but statistically significant negative impact on OS compared to the other predictors.

### D. Generic Satisfaction and Overall Satisfaction

A linear regression was used to test if GS significantly predicted OS. The overall regression was statistically significant ($R^2 = .617$, $F(7, 399) = 92.011$, $p = <.001$). GS variables were highly predictive of OS ($R^2 = .617$). The significant variables were EHR efficiency for patient work (GS1), Ease of use (GS4), and Support for clinical documentation (GS5), see Table III. Other GS variables were not significant.

TABLE III. REGRESSIONS PREDICTING OVERALL SATISFACTION (OS)

| Satisfaction | Regression statistics | | |
|---|---|---|---|
| | *Variable* | *β* | *p* |
| FS | FS1 | .459 | <.001 |
| | FS2 | .383 | <.001 |
| | FS3 | .332 | <.001 |
| | FS4 | -.151 | .003 |
| TS | TS1 | .276 | <.001 |
| | TS2 | .183 | .029 |
| | TS3 | .069 | .006 |
| GS | GS1 | .334 | <.001 |
| | GS4 | .282 | <.001 |
| | GS5 | .288 | <.001 |

TABLE IV. REGRESSIONS MODELL PREDICTING OVERALL SATISFACTION (OS)

| Satisfaction | Regression statistics | | | |
|---|---|---|---|---|
| | *β* | *Std. β* | *se* | *p* |
| GS1 | .261 | .262 | .041 | <.001 |
| GS4 | .190 | .197 | .038 | <.001 |
| GS5 | .198 | .177 | .044 | <.001 |
| FS1 | .201 | .147 | .054 | <.001 |
| FS3 | .133 | .114 | .048 | .006 |
| FS2 | .141 | .107 | .060 | .019 |
| TS1 | .067 | .066 | .034 | .049 |

### E. *Stepwise Regression Model for Overall Satisfaction*

Multiple linear regression was used to test if the significant TS, FS, and GS variables significantly predicted the OS. The overall regression was statistically significant ($R^2 = 0.682$, $F(7, 399 = 122.362, p = < .001)$. The final model incorporating significant TS, FS, and GS variables is presented in Table IV and Figure 1. FS4, TS2, and TS3 were insignificant and excluded from the final model. There are no major multicollinearity issues (VIF < 2.591).

## IV. DISCUSSION

The findings of this study show that key factors influencing overall satisfaction with the EHR, represented by factors related to improving system efficiency and usability (GS1, GS4), enhancing support for clinical documentation and workflow (GS5, FS1, FS3), developing better customization options (FS2), and improving training (TS1). Addressing these areas can lead to a more user-friendly, efficient, and adaptive EHR system.

The results of this study emphasize several critical factors that determine overall satisfaction with the Electronic Health Record (EHR) system. These factors include improvements in usability and system efficiency (GS1, GS4), support for clinical documentation and workflow (GS5, FS1, FS3), customization capabilities (FS2), and comprehensive training programs (TS1).

### A. *Implications for EHR System Improvement*

Maximizing EHR systems' effectiveness and enhancing user satisfaction will lead to enhanced usability and workflow efficiency. Streamlining workflow processes and optimizing documentation tools can significantly improve user satisfaction. Implementing intuitive navigation, reducing redundant data entry, and integrating Artificial Intelligence (AI) assisted automation for repetitive tasks can enhance efficiency, minimize cognitive burden, and allow healthcare professionals to focus more on patient care.

While EHR system training moderately impacts overall satisfaction in our study, improving training methodologies can lead to better user adoption and experience. Personalized, role-specific training modules, hands-on simulations, and continuous learning opportunities can help users adapt more

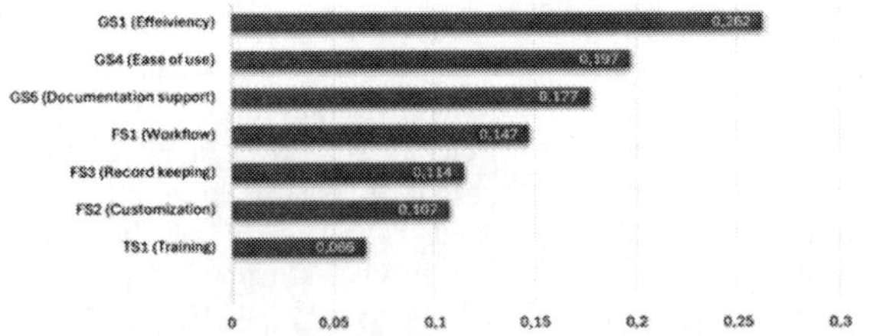


Figure 1. Predictor Importance in Overall Satisfaction (Standardized β).

effectively to system functionalities and leverage advanced features to improve productivity. Although training appears to have a minor effect in this study, it has been three years since a major upgrade of the EHR system was implemented. Training is a significant factor in user satisfaction with the implementation and major upgrades of EHR systems. Furthermore, training can be crucial for new EHR system users, regardless of when they start using it.

Providing greater flexibility in system configuration allows users to tailor the interface and workflows to align with their clinical routines. Offering customizable dashboards, adaptable templates, and user-defined shortcuts will enhance usability, reduce frustration, and improve overall engagement with the system.

Actively collecting system users feedback will ensure that the system evolves in response to user needs. Feedback could be done by incorporating user feedback through regular surveys, usability testing, and iterative design refinements. Establishing a feedback loop between clinicians and developers can drive enhancements that address real-world challenges and usability pain points.

By focusing on these areas, EHR systems can become more adaptive, user-centred, and conducive to efficient healthcare delivery, ultimately improving clinician satisfaction and patient outcomes.

The role of nonsignificant factors in EHR user satisfaction is important. While considerable attention has been given to factors that significantly impact overall satisfaction, it is equally important to consider factors that were not statistically significant. These elements still have the potential to improve satisfaction if they are better integrated or optimized within the EHR system. Possible reasons for nonsignificant factors could be: 1) Limited usage or awareness. Some features may not contribute to satisfaction simply because they are underutilized or not well-known among users; 2) Lack of adaptation to clinical workflow. If a feature is not intuitive or well-integrated, users may avoid it, leading to lower satisfaction scores; and 3) Technical limitations or interoperability issues. Poor implementation or lack of seamless integration with other systems can reduce the perceived value of a feature. It could be the cause of nonsignificant factors, such as Communication and patient interaction (FS4) and Care Coordination Support (GS7). EHR systems can enhance user engagement and functional satisfaction by addressing these gaps through usability improvements, workflow integration, and better user training, even for currently nonsignificant factors.

### B. Implications for Valkyrie Project

The findings from this study carry significant implications for the Valkyrie project, particularly in enhancing user experience and optimizing system functionality. Prioritizing usability and workflow efficiency should be a foundational goal in Valkyrie's development to ensure a seamless and intuitive interface that meets the needs of healthcare professionals. Variables such as GS5 (Support for Clinical Documentation) and FS1 (Workflow and Efficiency), both statistically significant predictors of overall satisfaction, directly inform key design goals of Valkyrie's VHR, such as seamless documentation and efficient data navigation. By aligning system development with these empirically identified needs, Valkyrie can proactively enhance clinician satisfaction and usability.

Valkyrie's success will rely on the support of comprehensive clinical documentation, enabling effortless and timely access to health data across systems. Ensuring interoperability and real-time data integration will improve decision-making and reduce administrative burdens.

The insights gained from this study can serve as a valuable framework for implementing user-centred design strategies to develop Valkyrie's VHR. Adaptive user interfaces, streamlined navigation, and intelligent data retrieval will enhance system usability and efficiency.

Additionally, pre- and post-implementation studies or similar evidence-based methodologies will be essential for evaluating Valkyrie's impact on user satisfaction. Continuous assessment and iterative improvements based on user feedback will contribute to a more responsive and effective system, ultimately leading to improved clinical workflows and patient care outcomes.

### C. Addressing Challenges in Accessing Large Journal Data

Healthcare professionals face significant challenges in efficiently accessing and interpreting large volumes of patient data stored in EHR systems. Data prioritization should be implemented to improve data usability and retrieval. This could be intelligent filtering mechanisms that dynamically highlight the most relevant clinical data by implementing context awareness. Implementing context-aware data prioritization could be an intelligent filtering mechanism that dynamically highlights the most relevant clinical data based on user roles, patient conditions, and clinical context. By leveraging real-time data analysis and AI-driven recommendations, the system can adapt to different care scenarios emphasizing acute conditions in emergency settings, chronic disease history in primary care, and clinical context. Context awareness ensures that clinicians receive the most relevant information when they need it, thereby reducing information overload and, at the same time, enhancing decision-making.

Further, in line with Valkyries VHR, a dynamic interface design could provide a user-friendly dashboard with structured overviews that categorize data by priority, timeline, and relevance. Customizable widgets and role-specific views should be integrated to improve the visualization of relevant data and thereby help to reduce cognitive overload.

Finally, focus on a context-aware optimized documentation retrieval process that transitions from free-text search dependency to structured data retrieval methods that use standardized terminologies and metadata indexing. Natural Language Processing (NLP) and semantic search functions should be incorporated to refine search accuracy and relevance, ensuring faster access to vital clinical documentation.

By integrating context-aware prioritization, role-specific interfaces, and advanced search capabilities, EHR systems can become more efficient and user-centred, improving clinical workflows and reducing the time spent navigating complex datasets.

## V. Conclusion and Future Work

This study highlights the key factors influencing user satisfaction with EHR systems, emphasizing the importance of usability, workflow efficiency, and clinical documentation support. Training was found to have a secondary but meaningful impact, especially during system transitions and for new users.

The findings provide several strategic recommendations for improving EHR systems, including the need for optimizing efficiency and usability by streamlining workflows and integrating intelligent automation, strengthening clinical documentation and workflow support to minimize administrative burdens, enhancing training strategies for better adoption and long-term satisfaction, providing flexible customization options to align with diverse clinical needs, and implementing continuous feedback mechanisms for iterative system improvements. These insights are directly applicable to the Valkyrie project, where user-centred design and seamless data accessibility will be crucial for the successful implementation of the VHR.

Future research should focus on longitudinal studies to assess how satisfaction evolves and how emerging technologies, such as AI-driven data filtering and real-time interoperability, can enhance EHR usability. Continuous evaluation and iterative improvements will ensure that EHR systems remain effective, responsive, and beneficial to healthcare professionals and patients. By applying these principles, existing EHR systems and new initiatives like the Valkyrie project can foster more intuitive, efficient, and user-centred digital healthcare environments, ultimately benefiting healthcare professionals and patient outcomes.

## References


[1] L. Poissant, J. Pereira, R. Tamblyn, and Y. Kawasumi, "The impact of electronic health records on time efficiency of physicians and nurses: a systematic review", J Am Med Inform Assoc., 12(5), pp. 505-16, Sep-Oct 2005, doi:10.1197/jamia.M1700.

[2] C. Tanner, D. Gans, J. White, R. Nath, and J. Pohl, "Electronic health records and patient safety: co-occurrence of early EHR implementation with patient safety practices in primary care settings", Appl Clin Inform., 11;6(1), pp. 136-47, Mar. 2015, doi:10.4338/ACI-2014-11-RA-0099.

[3] E. Li, J. Clarke, H. Ashrafian, A. Darzi, and A. Neves, "The Impact of Electronic Health Record Interoperability on Safety and Quality of Care in High-Income Countries: Systematic

Review", J Med Internet Res, 24(9):e38144, 2022, doi:10.2196/38144.

[4] J. L. Howe, K. T. Adams, A. Z. Hettinger, and R. M. Ratwani, "Electronic Health Record Usability Issues and Potential Contribution to Patient Harm", JAMA, 27;319(12), pp. 1276-1278, Mar. 2018, doi:10.1001/jama.2018.1171.

[5] K. Zheng, R. M. Ratwani, and J. Adler-Milstein, "Studying Workflow and Workarounds in Electronic Health Record-Supported Work to Improve Health System Performance", Ann Intern Med., 2;172(11 Suppl), pp. S116-S122, Jun. 2020, doi: 10.7326/M19-0871.

[6] T. Heponiemi et al., "Electronic Health Record Implementations and Insufficient Training Endanger Nurses' Well-being: Cross-sectional Survey Study", J Med Internet Res., 23;23(12):e27096, Dec. 2021, doi:10.2196/27096.

[7] T. G. Solvol, C. Granja, S. Cassidy, Ø. S. Solvang, and O. Lintvedt, "Valkyrie: A Distributed Service-Oriented Architecture for Coordinated Healthcare Services", pp. 42-46. International Conference on eHealth, Telemedicine, and Social Medicine (eTELEMED), 2023. Retrieved: April, 2025, https://hdl.handle.net/11250/3125119.

[8] Ministry of Health and Care Services, "One citizen – one Health Record", Whitepaper. no. 9 (2012-2013). "Én innbygger - én journal". *St.Meld. nr. 9* (2012–2013), 2012. Retrieved: April, 2025, https://www.regjeringen.no/no/dokumenter/meld-st-9-20122013/id708609/

[9] M. Quinn et al., "Electronic health records, communication, and data sharing: challenges and opportunities for improving the diagnostic process", Diagnosis (Berl). 27;6(3), pp. 241-248, Aug. 2019, doi:10.1515/dx-2018-0036.

[10] J.H. Holmes et al., "Why Is the Electronic Health Record So Challenging for Research and Clinical Care?", Methods Inf Med., 60(1-02), pp. 32-48, May 2021, doi:10.1055/s-0041-1731784.

[11] K. S. Williams and S. J. Grannis, "Patient-Centered Data Home: A Path Towards National Interoperability", Frontiers in Digital Health, 4, Jul. 2022, doi:10.3389/fdgth.2022.887015.

[12] O. Lintvedt, L. Marco-Ruiz, and R. Pedersen, "User Satisfaction with Recently Deployed Electronic Health Records", Studies in Health Technology and Informatics, vol. 302, pp. 192–196, May 2023, doi:10.3233/SHTI230101.

[13] R. J. Little, "A Test of Missing Completely at Random for Multivariate Data with Missing Values", Journal of the American Statistical Association, 83:404, pp. 1198-1202, Jul. 1988, doi: 10.1080/01621459.1988.10478722.

[14] O. Lintvedt, L. Marco-Ruiz, and R. Pedersen, "User Satisfaction with Recently Deployed Electronic Health Records", Stud Health Technol Inform., 18;302, pp. 192-196, May 2023, doi:10.3233/SHTI230101.

[15] O. Lintvedt, E. S. Nordheim, T. G. Solvoll, R. Pedersen, "Evaluating Usability and User Satisfaction in Hospital-Based EHR Systems: A Comparative Study of SUS and NuHISS in the Context of the Valkyrie Project", 2025, in press.

# Accuracy Evaluation of Computer Vision-based Markerless Human Pose Estimation for Measuring Shoulder Range of Motion

Thomas Hellstén and Jonny Karlsson
Arcada University of Applied Sciences
School of Engineering, Culture and Wellbeing, Helsinki, Finland
emails: thomas.hellsten@arcada.fi, jonny.karlsson@arcada.fi

*Abstract*— **TeleRehabilitation (TR) requires precise joint Range Of Motion (ROM) measurement methods. This study assessed the accuracy of a Computer Vision (CV)-based markerless Human Pose Estimation (HPE) application for active shoulder ROM by comparing it with Universal Goniometry (UG) in 20 healthy volunteers. The correlation coefficients between the two methods were 0.94 for shoulder extension, 0.83 for adduction, 0.76 for abduction, and 0.67 for flexion, with mean differences ranging from 3.6° in flexion to -7° in adduction. These findings indicate that the markerless CV application is a moderately accurate tool for measuring shoulder ROM.**

***Keywords-computer vision; markerless; range of motion; shoulder.***

## I. Introducttion

Measuring joint Range Of Motion (ROM) is essential for healthcare professionals to evaluate and treat patients with joint disorders, as it quantifies flexibility and function to establish baseline mobility, monitor progress, and tailor interventions for rehabilitation [1]. Traditionally, this evaluation is done by professionals with a Universal Goniometry (UG) at clinics [2]. ROM assessment methods are evolving with TeleRehabilitation (TR) approaches [3] that show the feasibility of CV-based Human Pose Estimation (HPE) for at-home use via a computing device with an integrated camera. This study builds on earlier testing of CV-based HPE for shoulder ROM [5] by assessing its accuracy in measuring active shoulder flexion, extension, abduction, and adduction compared to manually measured UG results, with the extended abstract structured into Section II (methods), Section III (accuracy results), Section IV (discussion), and Section V (conclusions and work).

## II. Methods

A prototype application was developed to evaluate the accuracy of CV-based marklerless HPE for shoulder ROM measurements. The application uses You Only Look Once, version 8 (YOLOv8) pose estimation to detect the shoulder, elbow, and hip joints and the law of cosines to calculate the shoulder angle [5]. Prior to the study, written consent was obtained from participants after the research protocol was approved by the Arcada University of Applied Science Research Committee (March 2024). Joint angles measured automatically by the CV-based markless HPE application were compared with manually measured UG results on 20 healthy participants (students and personnel aged 18 or older) by two near-graduating physiotherapy students using standardized assessments of the left shoulder (active flexion, extension, abduction, and adduction) with instructions provided from a test manual and demonstration by the test leader. All measurements were recorded in a blinded manner. The prototype registered joint angles with a timestamp that was concealed until manual UG measurements, and the results were collected into a single file for analysis. Environmental factors were standardized by performing all measurements in the same room with consistent lighting, temperature, and computer and webcam placement (55 cm high and approximately 2 m from the participant).

### A. *Statistical analysis*

Descriptive statistics for the voluntary test participants were computed and are reported in the results section. Accuracy was assessed with Pearson's correlation (*r*) analysis to calculate the correlation between the CV-based markerless HPE application and the manually UG measured shoulder ROMs. Bland–Altman plot analysis was used to estimate the agreement between the two methods. For correlation analysis, the following classification was used: 1.00–0.90 as very strong, 0.89–0.70 as strong, 0.69–0.50 as moderate, 0.49–0.30 as weak, and 0.29–0 as very weak [6].

## III. Results

The study participants (N=20) were healthy young adults (10 females, 20 males) aged between 20 and 33 years (mean: 22.9 years), as shown in Table 1.

TABLE I. Characteristics of the Study Participants

| **Participants (n[a])** | **Age, years; mean (SD[b])** | **Length; (cm[c]); mean (SD[b])** | **Weight (kg[d]); mean (SD[b])** | **BMI[e]; mean (SD[b])** |
|---|---|---|---|---|
| Total (20) | 23.4 (1.9) | 178.7 (7.9) | 78.0 (12.8) | 24.3 (2.3) |
| Female (4) | 23.5 (1.9) | 170.3 (3.0) | 64.3 (4.4) | 22.1 (0.8) |
| Male (16) | 23.1 (2.0) | 180.8 (7.3) | 81.4 (11.8) | 24.8 (2.3) |

a. n: number of participants, b. SD: standard deviation, c. cm: centimeter, d. kg: kilogram, e. BMI: body mass index (kg/m2)

The results showed a very strong correlation (Persons *r* value) in active extension 0.94, strong correlation in active adduction 0.83, abduction 0.76 and moderate correlation in active flexion 0.67. The mean difference (degrees) between the two methods was lowest in active shoulder flexion (3.6°). The highest mean difference (-7°) was in active shoulder adduction. Detailed results are shown in Figure 1.

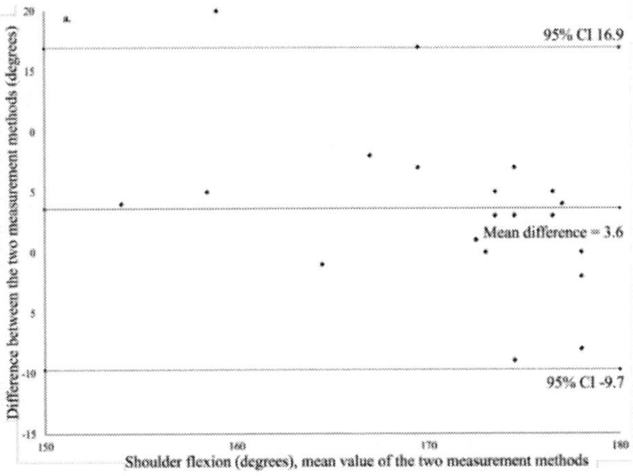


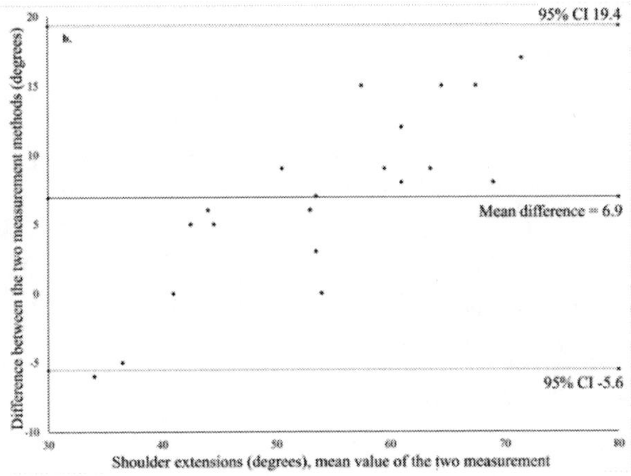


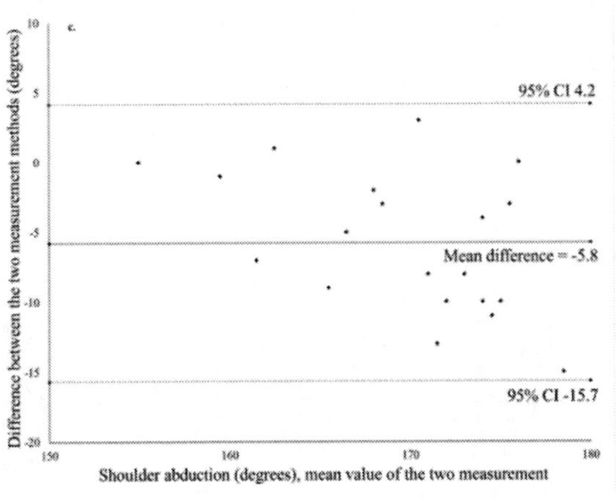


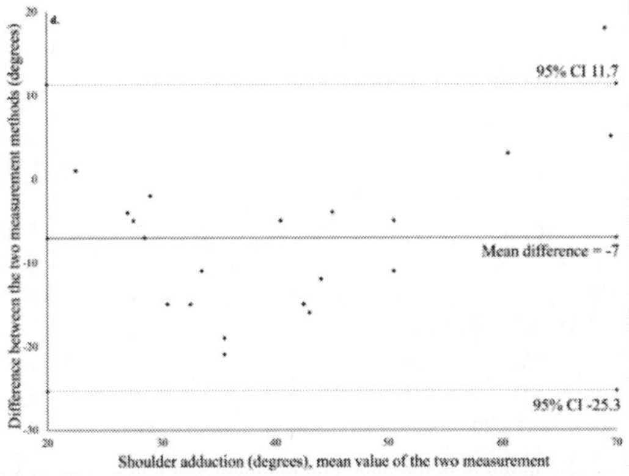


Figure 1. Bland–Altman plots for (a) shoulder flexion, (b) shoulder extension, (c) shoulder abduction, (d) shoulder adduction. The outer lines represent 95% limits of agreement. The middle line represents the mean of the differences between the two measurement methods.

Active shoulder flexion had two outliers outside the limits of agreement, and active shoulder adduction had one outlier outside the limits of agreement.

## IV. Discussion

The study shows a strong correlation between UG and the CV-based markerless HPE method for nearly all shoulder ROMs, extension, abduction, and adduction, with the exception of shoulder flexion. In abduction and adduction, the CV-based markerless HPE application tended to yield slightly lower values than UG. Despite some variation between the two methods, the results remain clinically acceptable given that UG, the standard for professionals, has its own measurement error [7]. However, since accuracy was assessed under standardized clinical conditions using a small group of healthy young adults, it is uncertain if these findings would be equally valid for older individuals or those with various diseases. Consequently, the prototype should be evaluated on more diverse populations and under different background and lighting conditions. Preliminary tests indicate that the prototype is sensitive to darker environments and inconsistent backgrounds, suggesting that adjustments to the underlying YOLOv8 pose estimation model may be needed for reliable home use. Overall, our results demonstrate that CV-based markerless HPE has great potential.

## V. Conclusion And Future Work

Our study indicates that CV-based markerless HPE applications show great potential as a promising means for implementing automatic real-time TR. However, they must be rigorously tested and developed collaboratively before becoming a standard tool in daily healthcare practice.

## Acknowledgment

This work was supported by the "Fonden för teknisk undervisning & forskning" (TUF) under Grant 1580, a fund for supporting education and technical research at Arcada University of Applied Sciences.

## References

[1] V. Lind, M. Svensson, and M. L. Harringe, "Reliability and validity of a digital goniometer for measuring knee joint range of motion," Measurement in Physical Education and Exercise Science, vol. 26, no. 3, pp. 191-198, Nov. 2022, doi: 10.1080/1091367X.2021.2004150.

[2] T. G. Russell, G. A. Jull, and R. Wootton, "Can the Internet be used as a medium to evaluate knee angle?," Manual Therapy, vol. 8, no. 4, pp. 242-246, Nov. 2003, doi: 10.1016/S1356-689X(03)00016-X..

[3] T. Hellstén, J. Arokoski, T. Sjögren, A.-M. Jäppinen, and J. Kettunen, "The Current State of Remote Physiotherapy in Finland: Cross-sectional Web-Based Questionnaire Study," JMIR Rehabil Assist Technol, vol. 9, no. 2, p. e35569, Jun. 2022, doi: 10.2196/35569.

[4] E. Hannink et al., "Validity and feasibility of remote measurement systems for functional movement and posture assessments in people with axial spondylarthritis," Healthcare Technology Letters, vol. 9, no. 6, pp. 110-118, Dec. 2022, doi: 10.1049/htl2.12038.

[5] T. Hellstén and J. Karlsson, "Using Computer Vision based Markerless Pose Estimation for Measuring Shoulder Range of Motion," presented at the eTELEMED 2024, The Sixteenth International Conference on eHealth, Telemedicine, and Social Medicine, Barcelona, Spain, 2024. [retrieved: February, 2025] https://www.thinkmind.org/library/eTELEMED/eTELEMED_2024/etelemed_2024_1_60_40016.html.

[6] D. E. Hinkle, S. G. Jurs, and W. Wiersma, Applied statistics for the behavioral sciences / Dennis E. Hinkle, William Wiersma, Stephen G. Jurs, 5th ed. Boston: Houghton Mifflin, 2003.

[7] M. J. Kolber and W. J. Hanney, "The reliability and concurrent validity of shoulder mobility measurements using a digital inclinometer and goniometer: a technical report,", Int J Sports Phys Ther, vol. 7, no. 3, pp. 306-13, June 2012.

# User Perspectives on Electronic Health Record Functionality

## A Qualitative Evaluation of Clinical Experiences

[a] Espen Solbakken Nordheim[1], [b] Ove Lintvedt[1,2], [c] Rune Pedersen[1,2]

[1]Norwegian Centre for E-health Research,
University Hospital of North Norway,
Tromsø, Norway
[2]Norway Faculty of Nursing and Health Sciences
Nord University
Bodø, Norway

e-mail: [a]espen.solbakken.nordheim@ehealthresearch.no, [b]ove.lintvedt@ehealthresearch.no,
[c]rune.pedersen@ehealthresearch.no

*Abstract*— **This study examines the experiences of clinical users with one Electronic Health Record (EHR) system, DIPS Arena. In a survey for users of DIPS Arena across several healthcare institutions in Northern Norway, 57 respondents gave text-based feedback in the functionality of the EHR system. The research identifies key challenges through a thematic analysis based on text answers in a survey. Four main themes emerged: system integration and stability, usability and design, administrative and clinical functionality, and training and support. The findings reveal that some health care professionals had problems with the EHR system in use, including fragmented data sharing, poor user interface design, increased administrative burdens, and inadequate training programs. These challenges could impact on workflow efficiency and patient care quality, highlighting the need for continuous improvements and user feedback from the EHR producer. The study concludes that while this feedback is not representative as a survey, it adds to the knowledge on what EHR producers need to be aware of when improving their product.**

***Keywords-Electronic Health Record (EHR); Functionality; User Satisfaction; Organization and technology; Installed Base.***

## I. INTRODUCTION

Over the past decade, the Norwegian government, as many western countries [1], has embarked on a comprehensive initiative to modernize its Electronic Health Record (EHR) infrastructure. EHRs offer numerous benefits, including improved efficiency, better access to patient information, enhanced communication among healthcare providers, and higher quality and safety of patient care [2]. However, potential adverse effects have also been reported, such as lack of integration between systems [3][4], poor usability [5], time-consuming documentation procedures [6], clinician burnout [7], and insufficient training [2].

These broader trends and challenges are also evident in the Norwegian healthcare system, where national efforts to modernize EHRs have led to the adoption of different implementation strategies across regions. The Central Regional Health Authority adopted a "big bang" strategy, implementing the EPIC system simultaneously across multiple institutions, which encountered notable challenges and difficulties [8][9]. In contrast, the other three Regional Health Authorities pursued an evolutionary implementation strategy, introducing the DIPS Arena system by gradually building upon the existing infrastructure, which emphasizes the importance of developing new systems by leveraging and integrating with existing technologies to minimize disruption and enhance adoption [10].

Despite the differences in implementation, the effectiveness of any EHR system ultimately depends on how well it supports clinical work in practice. Previous evaluations of DIPS Arena have indicated overall user satisfaction with the system [11]. Previous studies looking at both DIPS Arena and its predecessor DIPS Classic have identified significant shortcomings in critical functionalities, like clinical workflow support and medication management [12][13]. Such gaps highlight ongoing challenges in achieving the full potential of EHR systems and underscore the need for deeper, qualitative insights into users' real-world experiences.

To address this gap, the present study analyzes open-ended survey responses to gain deeper insight into users' experiences with the EHR system. By examining how clinicians describe its functionality in their own words, the study aims to identify recurring themes, highlight critical areas for improvement, and assess how such insights might contribute to enhancing overall user experience.

The overall structure of this paper includes five sections: Section II explains the qualitative methods used; Section III presents the results; Section IV discusses the findings; and Section V concludes the study.

## II. METHODOLOGY

### A. *Setting of study*

A quantitative study was conducted in 2024 among clinical Electronic Health Record (EHR) users of DIPS Arena, three years after the systems' implementation.

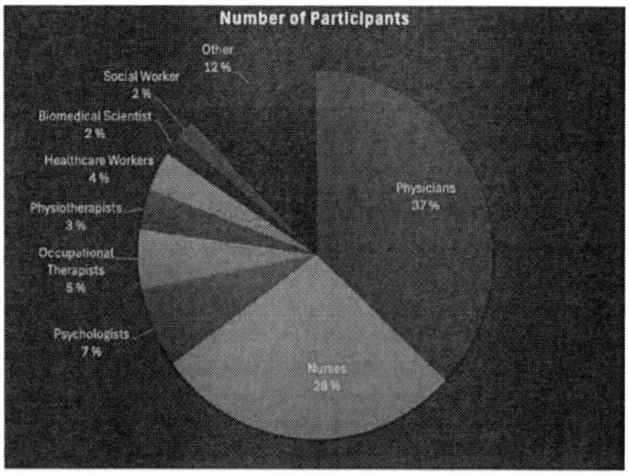


Figure 1. Number of participants.

A total of 549 participants were recruited via email from several healthcare institutions within the Northern Regional Health Authorities to ensure a representative sample of clinical users. The sample included participants from a diverse range of occupations, as seen in Figure 1. Prior to the main study, the survey instrument was pilot-tested with a subset of eight clinical users to assess face validity and ensure that the questions were clearly understood.

### *B. Data collection*

The survey comprised both closed-ended and open-ended items. The closed-ended questions asked participants to rate specific functionalities in the EHR system using a five-point Likert scale. In addition, an open-ended free text question was included to capture qualitative insights regarding the systems' functionality. Participants were asked if there was anything else they wanted to comment on beyond the functionalities mentioned. A total of 57 users provided text-based responses to this question. Among them, physicians formed the largest group, with 21 individuals, followed by nurses, who accounted for 16 respondents. Psychologists were represented by 4 participants, while occupational therapists numbered 3. Physiotherapists and healthcare workers each contributed 2 participants. Additionally, there was 1 biomedical scientist and 1 social worker among the respondents. Finally, 7 individuals identified their occupation as "Other," reflecting a category of diverse or unspecified roles.

### *C. Analysis*

The data were analyzed using thematic analysis, following Braun and Clarkes' [14] framework. The analysis was conducted using Microsoft Word 365 (Microsoft Corporation). The first author (ESN) initially coded the text-based responses, and a total of 72 codes were developed. RP reviewed the codes, and any discrepancies were resolved through discussion until consensus was reached. These codes were then grouped into conceptually similar categories through an iterative process of comparison and refinement.

TABLE I. CODING AND THEMES

| Theme | Qualitative Themes | |
|---|---|---|
| | ***Number of codes*** | ***Example Codes*** |
| System integration, technical stability, and communication | 17 | Lack of system integration, critical info not integrated across systems, no integration between CP3 and DIPS, systems load slowly |
| Usability, design, and navigation | 10 | Illogical shortcut keys, poor screen layout, too much clicking |
| Administrative processes, documentation, and clinical functions | 17 | Double registration in care pathways, manual medication entry, increased documentation workload |
| Training, implementation, and support | 2 | Poor training, inadequate e-learning and transition support |

A total of 46 codes were retained for further analysis, as they were considered relevant to the aim of the study. The identified codes were subsequently organized into overarching themes that reflected the experiences of clinical users.

The themes were reviewed for clarity, then clearly defined and named to capture the underlying meaning of the codes. The final structure consisted of four overarching themes: (1) system integration, technical stability, and communication (17 codes), highlighting issues with interoperability, performance, and messaging; (2) usability, design, and navigation (10 codes), capturing interface and customization problems; (3) administrative processes, documentation, and clinical functions (17 codes), covering inefficiencies in data entry, documentation, and medication workflows; and (4) training, implementation, and support (2 codes), reflecting limited user preparation and transition support. Each theme was supported by representative quotations to ensure alignment with user perspectives. Table 1 shows examples of coding. The final themes were refined and validated through discussion between ESN and RP.

### *D. Ethics*

All participants provided informed consent. The responses were anonymized and stored securely to ensure the confidentiality of participants' data throughout the study. The Regional Committee for Medical and Health Research Ethics in Northern Norway was consulted. According to national regulations and ethical guidelines, formal approval was not required because the study did not involve biomedical research and all data was anonymized.

## III. RESULTS

### *A. Theme 1: System integration, technical stability and communication*

The first theme emerged around issues related to the integration of different Information Technology (IT) solutions, system stability, and digital communication functions. Respondents repeatedly mentioned that the lack of seamless integration among different systems used by the hospitals (DIPS, Metavisjon, and other solutions) leads to fragmented data sharing and delays in patient care. Users reported that data does not flow automatically between systems, resulting in additional manual work and potential risks to patient safety. For example, one bioengineer commented, "Generally speaking, we need a data system that communicates with the rest of the country. We in Tromsø do not see [data from] the rest of the country," emphasizing the need for a unified and interconnected data system.

Many respondents described experiencing frequent log-in and log-out problems, which were compounded by slow system performance. This not only delayed clinical workflows but also forced users to manually verify and re-enter data across different platforms. A nurse remarked on the issue of critical information being split across separate systems: "Critical information in DIPS Arena and the National Patient Portal must be merged as soon as possible. The synchronization should be completely automatic and without the need to approve the data transfer. This will safeguard patient safety in the best possible way."

Respondents also discussed inadequate digital messaging solutions, such as the inability to send messages to municipalities from certain outpatient clinics, further widening the communication gap between departments. The overall sentiment indicated that the systems' technical instability and lack of proper integration not only contributed to workflow inefficiencies but also caused considerable frustration among clinical staff.

### B. *Theme 2: Usability, design and navigation*

The second theme focused on challenges associated with the systems user interface and design elements. Many users expressed their dissatisfaction with the logical flow of menus and the overall layout of the system. Complaints were centered on the unstructured design, inadequate screen utilization, and inefficient navigation mechanisms that require numerous unnecessary clicks. One physician compared the current system unfavorably with its predecessor, stating, "User-friendliness in DIPS Arena is poorer, and less organized than DIPS Classic," highlighting that the new system has not enhanced, but rather detracted from, the user experience.

Additional concerns related to the inability to personalize or customize the interface were also raised. The absence of customizable settings forced users to adapt to a rigid interface that did not align with their specific workflow needs. This issue was underscored by a clinician's observation regarding the inconsistency of shortcut keys: "Keyboard shortcuts!! I use/have used keyboard shortcuts for all software – but the DIPS-Arena setup is unusable. Keyboard shortcuts that one has become accustomed to change function between versions. 'CTRL-D': first 'insert diagnosis,' now 'delete document'!!!!" Such changes in key functions contributed to the perception that the system design was not only counterintuitive but also disruptive.

Furthermore, users wrote that the placement of symbols and buttons across the interface was disorganized. A nurse highlighted that "Very messy design. The icons are not intuitive and are placed in different positions," indicating that the scattered placement of icons and the requirement to frequently move the mouse across different areas of the screen hindered efficient data entry and increased cognitive load. The design issues were also linked to a potential compromise in patient safety, as one physician noted that the burdensome interface resulted in more manual corrections and delays in retrieving vital patient information.

### C. *Theme 3: Administrative processes, documentation and clinical functions*

The third theme addressed the burdens associated with administrative tasks, documentation, and core clinical functions within the Electronic Health Record (EHR) system. Many respondents voiced concerns about redundant administrative procedures, particularly the necessity for double data entry and the use of poorly designed journal templates. Several users described the workflow as time-consuming, with one occupational therapist lamenting, "I feel like quitting my job every time I have to do this completely pointless and burdensome registration." Such comments point to the inefficiency embedded in the administrative aspects of the system, where manual inputs and repetitive documentation tasks detract from time available for direct patient care.

The issue of double registration was mentioned repeatedly, with clinicians noting that the lack of automation in critical areas, such as medical coding and other standardized templates, resulted in unnecessary administrative burden. One physician remarked that "The so-called savings with EHR have contributed to secretaries being laid off and tasks being transferred to the doctors," a shift that not only increased the workload for physicians but also compromised the overall efficiency of the healthcare delivery system.

Another area of concern was the handling of referrals and other administrative documents. In some departments received referrals were misdirected, forcing staff to spend considerable time sorting and redirecting them. One respondent reported that this misrouting "…steals about seven hours of work time per week from our outpatient clinic," thereby highlighting the inefficiencies introduced by the current system design.

Furthermore, challenges with medication management were also brought to the forefront. A healthcare professional noted, "Patients' medications are not digitized in our system. This must be implemented," emphasizing the need for an integrated digital solution for medication tracking. The lack of efficient documentation protocols and standard templates was seen to further impede the effective recording and retrieval of clinical data.

*D. Theme 4: Training, implementation and support*

The final theme emerged from respondents' experiences with the training, implementation, and support provided during the transition from the previous system (DIPS Classic) to the updated version (DIPS Arena). Even though there was a small number of codes, there were some users expressing frustration over the inadequate e-learning modules and classroom sessions. One informant encapsulated the sentiment by stating, "The implementation process transitioning from Classic to Arena was too challenging, and we had no direct contact with the system developers." This lack of direct access to developers during the critical phase of system implementation appears to have significantly contributed to user stress and misinterpretations of system functionalities.

One respondent criticized the training materials, noting that the content was neither comprehensive nor tailored to the specific needs of different user groups commenting "The training component, which consisted of e-learning and classroom instruction on a screen, showed so little of what was supposed to be learned that one could hardly see anything. It should have been arranged so that you could choose which parts of the program you needed, as less than half was relevant to my job." This sentiment underscores the mismatch between training content and the practical needs of clinical staff, which in turn led to increased reliance on self-directed learning and peer support.

## IV. Discussion

Overall, the findings across these four themes reveal challenges related to integration, design, administrative functionality, and training support. The responses from clinical users illustrate that the Electronic Health Record (EHR) system DIPS has been accompanied by operational disruptions, affecting both the efficiency of clinical workflows and the overall quality of patient care.

The first theme, concerning system integration, technical stability, and communication, reveals a fragmented digital infrastructure in which different systems—such as DIPS, Metavisjon, and various ambulance solutions, fail to communicate seamlessly. This fragmentation is not a novel observation in the literature; several studies have noted that a lack of interoperability often results in manual workarounds and increased risk to patient safety [3][4]. Our respondents' emphasis on the need for a system that "talks with the rest of the country" reinforces the call for more unified, interoperable systems. Moreover, the technical instability and slow system performance reported by users are consistent with previous observations that EHR transitions can disrupt clinical workflows [6].

The second theme focuses on usability, design, and navigation. Users reported issues with the user interface, including poorly structured menus, non-intuitive iconography, and inconsistent shortcut keys. These findings align with existing studies that argue poor usability and design can increase cognitive load and even contribute to clinician burnout [7][9]. The comparison made by one physician between DIPS Arena and DIPS Classic indicates that the redesign did not meet some users' expectations in terms of efficiency and intuitiveness. The physician stated that the interface in DIPS Arena was inferior to DIPS Classic. It is noteworthy that this argument is presented three years after implementation.

The third theme, administrative processes, documentation, and clinical functions, reflects a critical concern regarding increased administrative workload. Respondents reported double registrations, burdensome journal templates, and an inefficient referral management system. This aligns with previous research linking increased documentation requirements to reduced patient interaction time and greater clinician frustration [6][8]. The observation that "The so-called savings with EHR have contributed to secretaries being laid off and tasks being transferred to the doctors" underscores the frustration among some clinicians. This finding supports the notion that while EHR systems are designed to enhance operational efficiency, their practical implementation can sometimes lead to counterproductive outcomes.

The final theme, concerning training, implementation, and support, further suggest that the transition from DIPS Classic to DIPS Arena was not sufficiently supported by training programs or direct engagement with system developers for personalization. Moreover, these issues appear to have persisted over time. The lack of adequate training has been highlighted in previous studies as a key factor that hinders successful system adoption [2].

Our findings align closely with previous studies on this system, which identified significant functionalities—such as drug treatment overview, prescribing, and care planning [12], as well as medication management, clinical workflow support, and system stability, as areas associated with lower user satisfaction [13]. Together, these studies underscore that the challenges inherent in the DIPS Arena system are not new, reinforcing the need for continuous user feedback and iterative improvements to better support the evolving demands of health information infrastructures.

Despite offering valuable insights into user experiences with the DIPS Arena system, this study has some limitations. The relatively small subset of qualitative responses (57 participants) may not capture the full spectrum of user experiences, and the reliance on self-reported data introduces potential biases. Respondents were likely more inclined to provide negative feedback, potentially due to self-selection bias, in which individuals with adverse experiences felt more compelled to respond—thereby limiting the representation of positive perspectives.

## V. Conclusion and Future Work

In this paper, we set out to explore and understand clinical users' experiences with the DIPS Arena Electronic Health Record (EHR) system, focusing particularly on its functionality and impact on clinical workflows. Our findings reveal that while the DIPS Arena system was developed with the intention of modernizing clinical workflows and enhancing patient safety, our findings reveal ongoing challenges in system integration, usability, administrative processes, and support structures three years after implementation. This finding is not confirmed by the quantitative results but still holds value, as it provides insight into user complexities. These issues, consistent with other studies, underscore the need for continuous user feedback and iterative refinements to better align the system with real-world clinical needs. Future research with more complete samples, combining qualitative and quantitative data, is needed to further elucidate these challenges.

## REFERENCES


[1] M. Hertzum, G. Ellingsen, and Å. Cajander, "Implementing large-scale electronic health records: experiences from implementations of Epic in Denmark and Finland," International journal of medical informatics, vol. 167, p. 104868, 2022.

[2] C. H. Tsai et al., "Effects of electronic health record implementation and barriers to adoption and use: a scoping review and qualitative analysis of the content," Life, vol. 10, no. 12, p. 327, 2020.

[3] E. Li, J. Clarke, H. Ashrafian, A. Darzi, and A. L. Neves, "The impact of electronic health record interoperability on safety and quality of care in high-income countries: systematic review," Journal of medical Internet research, vol. 24, no. 9, p. e38144, 2022.

[4] E. Li et al., "Physician experiences of electronic health records interoperability and its practical impact on care delivery in the English NHS: A cross-sectional survey study," medRxiv, p. 2024.07. 25.24311018, 2024.

[5] M. Cahill, B. J. Cleary, and S. Cullinan, "The influence of electronic health record design on usability and medication safety: systematic review," BMC Health Services Research, vol. 25, no. 1, p. 31, 2025.

[6] K. Malm-Nicolaisen, A. J. Fagerlund, and R. Pedersen, "How Do Users of Modern EHR Perceive the Usability, User Resistance and Productivity Five Years or More After Implementation?," in MEDINFO 2021: One World, One Health–Global Partnership for Digital Innovation: IOS Press, 2022, pp. 829-833.

[7] C. P. West, L. N. Dyrbye, and T. D. Shanafelt, "Physician burnout: contributors, consequences and solutions," Journal of internal medicine, vol. 283, no. 6, pp. 516-529, 2018.

[8] G. Ellingsen, M. Hertzum, and L. Melby, "The tension between national and local concerns in preparing for large-scale generic systems in healthcare," Computer Supported Cooperative Work (CSCW), vol. 31, no. 3, pp. 411-441, 2022.

[9] S. Lohmann-Lafrenz et al., "Usability of an electronic health record 6 months post go-live and its association with burnout, insomnia and turnover intention: a cross-sectional study in a hospital setting," BMJ Health & Care Informatics, vol. 32, no. 1, p. e101200, 2025.

[10] R. Pedersen, E. S. Nordheim, O. Lintvedt, A. J. Fagerlund, G-H. Severinsen and K. Malm-Nicolaisen, "The Knowledge of Implementation Strategies: Impact of the Installed Base," Studies in Health Technology and Informatics, vol. 305, pp. 273-276, 2023.

[11] O. Lintvedt, E. Nordheim, and R. Pedersen, "Electronic Health Records User Satisfaction", Proceedings of The Fifteenth International Conference on eHealth, Telemedicine, and Social Medicine (eTELEMED), pp. 52-57, 2023.

[12] O. K. Lintvedt et al., "Electronic Health Records User Experiences: a Nationwide Survey From Norwegian Hospitals," Proceedings of The Fourteenth International Conference on eHealth, Telemedicine, and Social Medicine (eTELEMED), pp. 52-57, 2022.

[13] T. R. Schopf, B. Nedrebø, K. O. Hufthammer, I. K. Daphu, and H. Lærum, "How well is the electronic health record supporting the clinical tasks of hospital physicians? A survey of physicians at three Norwegian hospitals," BMC health services research, vol. 19, pp. 1-9, 2019.

[14] V. Clarke and V. Braun, "Thematic analysis," The journal of positive psychology, vol. 12, no. 3, pp. 297-298, 2017.

# Advancing the Management of Paediatric Growth Hormone Deficiency

## The 4 W's and the Role of Digital Health and Patient Support

Paul Dimitri
Department of Paediatric Endocrinology
Sheffield Children's NHS Foundation Trust
Sheffield, United Kingdom
e-mail: paul.dimitri@nhs.net

Indraneel Banerjee
Department of Paediatric Endocrinology
Royal Manchester Children's Hospital
Manchester, United Kingdom
e-mail: Indi.Banerjee@mft.nhs.uk

Marta Ciaccio
Department of Endocrinology
Hospital de Pediatría S.A.M.I.C.
Buenos Aires, Argentina
e-mail: martaciaccioar@yahoo.com.ar

Antonio De Arriba
Pediatric Endocrinology Unit
Hospital Universitario Materno-Infantil Miguel Servet
Zaragoza, Spain
e-mail: adearriba@salud.aragon.es

Sandro Loche
Research Unit for Innovative Therapies in Endocrinopathies
Bambino Gesù Children's Hospital, IRCCS
Rome, Italy
e-mail: sandroloche@gmail.com

Azriyanti Anuar Zaini
Department of Paediatrics
University of Malaya
Kuala Lumpur, Malaysia
e-mail: azriyanti@ummc.edu.my

Abel López-Bermejo
Pediatric Endocrinology Research Group
Dr. Josep Trueta Hospital
Girona, Spain
e-mail: alopezbermejo@idibgi.org

In-Hyuk Jung
Department of Paediatrics National Health Insurance Service
Ilsan Hospital
Goyang-si, Korea
e-mail: medicaluke@naver.com

Bassam Saleh Bin-Abbas
Department of Pediatrics Endocrinology
King Faisal Specialist Hospital and Research Center
Riyadh, Saudi Arabia
e-mail: benabbas@kfshrc.edu.sa

Edoardo Vergani
Department of Translational Medicine and Surgery
IRCCS – Catholic University of the Sacred Heart
Rome, Italy
e-mail: edoardo.vergani@outlook.it

Ekaterina Koledova
Global Medical Affairs Cardiometabolic & Endocrinology
Merck Healthcare KGaA
Darmstadt, Germany
e-mail: ekaterina.koledova@merckgroup.com

***Abstract* - Optimizing care for children with Growth Hormone Deficiency (GHD) is a priority in paediatric endocrinology. This article summarizes insights from an expert advisory board that explored the "4 W's" of GHD management: who needs the most support, what issues are critical, when to intervene, and why it matters. Key recommendations included early intervention, personalized support programs, and the use of digital tools to enhance adherence and outcomes. The board also highlighted the need for further research on long-term impacts, Artificial Intelligence (AI)-driven analytics, and remote monitoring, including psychosocial considerations.**

***Keywords - adherence monitoring; digital health; growth hormone treatment; patient engagement; paediatric endocrinology; patient support programs.***

## I. Introduction

Paediatric Growth Hormone Deficiency (GHD) is a condition characterized by inadequate secretion of Growth Hormone (GH) from the pituitary gland, leading to impaired growth and development in children. It affects approximately 1 in 4,000 to 1 in 10,000 children worldwide [1]. GHD results from congenital or acquired causes, including genetic mutations, brain injuries, or tumours [2]. Left untreated, GHD can lead to significant health issues, including short stature, delayed puberty, and metabolic abnormalities [1][3].

Early diagnosis and intervention are crucial in managing GHD to ensure optimal growth, development and metabolic outcomes [4]. Treatment typically involves regular administration of recombinant human Growth Hormone (rhGH) injections starting in childhood, continuing to adolescence and in some cases into adulthood [5][6].

This multi-year treatment journey of a paediatric patient involves daily, or more recently - for a small number of patients, weekly, recombinant human GH (rhGH) injections [5]. The patient journey is fraught with challenges as each child and his/her caregivers adapt to this new reality, initiate treatment, and learn how to administer injections [5][7]. Challenges such as poor adherence to treatment regimens, needle phobia, and limited access to specialized care can impede effective management [5][8]–[11].

Along the way, the clinic nurse and/or Patient Support Programs (PSPs), in collaboration with the clinician, offer training and support [10]. Patients make regular visits to the clinic to monitor growth, assess metabolic parameters and make dose adjustments, as necessary. As children grow and develop during this journey of growth hormone therapy, there may also be emotional and behavioural challenges that need to be addressed. These issues may become particularly challenging during teenage years, as young people approach their move to adult endocrine care (transition). Staying informed, remaining adherent to therapy, and addressing challenges that occur along the way, are essential elements of ensuring optimal outcomes. This requires a multidisciplinary team involving physicians, clinic nurses, patient support team, and the child as well as his/her caregivers [10].

Over the last decade, digital health solutions have supported patients, caregivers, and their HealthCare Professionals (HCPs). They offer tools to monitor, engage and, in some cases, treat patients and their symptoms [12][13]. They enable HCPs to track and treat patients remotely more regularly, offering valuable real-world evidence and insights that help personalize and optimize treatment outcomes [14]. For paediatric GHD, digital health solutions, to primarily improve patient adherence, have been developed and made available to patients, caregivers and HCPs to support them along their journey [14][15]. These tools have been used by tens of thousands of patients and their HCPs across 40+ countries [14][15]. Real world evidence tracking millions of daily injections have enabled greater insights into the optimization of adherence and better growth outcomes.

Given this context we convened a meeting of ten expert physicians in paediatric endocrinology from both public and private healthcare institutions from across Europe, Middle East, Asia and Latin America selected based on their experience in using digital health technologies and patient-centric support, to look at "4 W's" in advancing GHD care. Specifically, we wanted to consider 'who' amongst our patients needs more support, 'what' issues need to be addressed, 'when' and how to intervene, and 'why' such interventions make a difference. These questions were intended to support HCPs in understanding and addressing key concerns during the patient journey – thereby developing bespoke and targeted management and care and ensuring optimal outcomes.

In this article, we summarize our discussions and offer a roadmap for further enhancing patient support using digital health technologies to address these "4 W's" in the management of GHD.

## II. OUR APPROACH

This paper is based on discussions from a meeting of ten expert physicians in paediatric endocrinology from across Europe, the Middle East, Asia, and Latin America. These experts were selected based on their extensive experience in using digital health technologies and patient-centric support in the management of paediatric GHD. The advisory board aimed to define a roadmap for enhancing patient support by addressing the "4 W's" in GHD care. The discussions focused on:

- Who Needs the Most Support? Identifying patient subgroups with the greatest need for additional support.
- What Are the Most Pressing Issues? Determining the key challenges in GHD management that need to be addressed.
- When & How Should We Intervene? Exploring optimal timing and methods for interventions.
- Why Do Such Interventions Make a Difference? Evaluating the impact and benefits of these interventions.

The experts explored how digital health solutions and patient support programs can address these questions to improve the patient journey and treatment outcomes.

## III. RESULTS

The advisory board's discussions yielded several key findings:

- Who Needs the Most Support: The experts identified several patient subgroups with heightened support needs. These include newly diagnosed children (struggling with treatment initiation), adolescents (facing challenges with self-injection, behavioural changes, and financial barriers), children from separated families (requiring coordination of injections), patients with multiple hormone deficiencies, and those transitioning to adult care (at risk of treatment discontinuation).
- What Are the Most Pressing Issues: The most pressing issues revolve around treatment adherence. This includes optimizing rhGH dosage, educating patients and caregivers, addressing psychosocial aspects, and managing potential side effects. The effective integration of digital tools into clinical workflows, without overburdening HCPs or patients, was also identified as a key challenge.
- When & How Should We Intervene: The advisory board emphasized that early and personalized interventions are critical. This includes timely growth tracking, IGF1 monitoring, and psychosocial assessments. Digital tools can facilitate early habit formation, provide personalized feedback, and enhance communication between families and support services. The first months of treatment are crucial for establishing long-term adherence.

- Why Do Such Interventions Make a Difference: The experts highlighted the importance of measuring the impact of interventions. AI and real-world data can be leveraged to optimize treatment, predict non-adherence, and personalize patient care [16]. Factors such as consistent injection timing, tailored digital support, and addressing patient and caregiver needs, beliefs, and cultural values can significantly improve outcomes.

The 4 W's and the role of digital health technologies and patient support are summarized in Figure 1.

## IV. Discussion

The advisory board emphasized that certain patient groups require more support, including newly diagnosed children, adolescents, children from separated families, patients with multiple hormone deficiencies and those transitioning to adult care.

The most pressing issue in managing paediatric GHD is ensuring treatment adherence. Non-adherence can stem from various factors, including injection fear, treatment fatigue, lack of family engagement, and difficulties integrating digital health solutions into care. Digital tools offer a promising avenue for supporting adherence, but they must be implemented strategically to avoid overburdening patients, caregivers, and HCPs. Educating patients and caregivers about the benefits of treatment and addressing psychosocial factors are also crucial for improving adherence.

Early and personalized interventions can significantly improve outcomes. Regular growth tracking, IGF-1 monitoring, and the use of digital reminders can help improve adherence. Digital tools can also provide personalized feedback, encourage early habit formation, and facilitate caregiver engagement. In particular, the first few months of treatment are critical for establishing long-term adherence.

Digital health tools, when combined with PSPs, can enhance patient support and engagement. AI and real-world data can be used to optimize treatment, predict non-adherence, and personalize patient care. Studies have shown that factors such as consistent injection timing, adjusting injection techniques, and providing tailored digital support can improve both adherence and growth outcomes. Personalized patient support programs, which use digital tools to target patients at risk of low adherence, have also demonstrated success.

## V. Conclusion And Future Directions

The advisory board concluded that a patient-centric approach, integrating digital health tools with personalized support, is essential for improving adherence and health outcomes in paediatric GHD patients. Early intervention, targeted support for high-risk groups, and leveraging AI and real-world data can enhance the effectiveness of GHD management.

The advisory board identified several key areas for future research and development:

- Further research is needed to evaluate the long-term impact of digital health interventions on growth outcomes and quality of life.
- There is a need for more personalized and integrated digital health solutions tailored to the specific needs of individual patients and their families.
- The role of AI in optimizing treatment decisions and predicting individual responses to therapy should be explored.
- Strategies to ensure equitable access to digital health technologies for all patients with GHD are needed.
- Further consideration of remote health monitoring tools, including assessments of psychosocial aspects of patient lives is needed as part of advancing care in GHD.

## Disclaimer

The data for this manuscript was based on an advisory board assembled and funded by Merck Healthcare KGaA.

## Conflicts of Interst

EK is an employee of Merck KGaA, Darmstadt, Germany and holds shares in the company. All other authors declare that their participation in the advisory board was sponsored through funding from Merck (CrossRef Funder ID: 10.13039/100009945).

## Acknowledgment

Medical writing assistance was provided by Merat Bagha and Dr. Kalika Mathur of Tiba Medical Inc., Beaverton, Oregon, USA.

## References

[1] T. Stanley, "Diagnosis of Growth Hormone Deficiency in Childhood." Curr. Opin. Endocrinol. Diabetes Obes., vol. 19, pp. 47-52, Feb. 2012, doi: 10.1097/MED.0b013e32834ec952.

[2] Endocrine Society, "Endocrine Society Patient Resources" 2022, Growth Hormone Deficiency, Available from: https://www.endocrine.org/patient-engagement/endocrine-library/growth-hormone-deficiency.

[3] D. Capalbo et al., "Update on early cardiovascular and metabolic risk factors in children and adolescents affected with growth hormone deficiency" Minerva Endocrinol., vol. 37, pp. 379-389, Dec. 2012. Available from: https://www.minervamedica.it/en/journals/minerva-endocrinology/article.php?cod=R07Y2012N04A0379.

[4] S. Stagi et al., "Management of Neonatal Isolated and Combined Growth Hormone Deficiency: Current Status." Int. J. Mol. Sci., vol. 24, pp.10114, Jun. 2023, doi: 10.3390/ijms241210114.

[5] M. O. Savage et al., "Adherence to r-hGH Therapy in Pediatric Growth Hormone Deficiency: Current Perspectives on How Patient-Generated Data Will Transform r-hGH Treatment Towards Integrated Care." Patient Prefer

Adherence, vol. 16, pp. 1663-1671, Jul. 2022, doi: 10.2147/PPA.S271453.

[6] R. S. Farber and J. R. Kerrigan, "The multiple indications for growth hormone treatment of pediatric patients." Pediatr. Ann., vol. 35, pp. 926-932, Dec. 2006, doi: 10.3928/0090-4481-20061201-06.

[7] S. Malik, C. Moloney, E. Koledova, J. Reston, and J. Weinman, "Designing a Personalized Digital Patient Support Program for Patients Treated With Growth Hormone: Key Design Considerations." J. Med. Internet Res., vol. 22, pp. e18157, Jul. 2020, doi: 10.2196/18157.

[8] A. M. Pereira et al., "Efficacy of needle-free administration of recombinant human growth hormone in adults with growth hormone deficiency." Br. J. Clin. Pharmacol., vol. 61, pp. 451-455, Apr. 2006, doi: 10.1111/j.1365-2125.2006.02585.x.

[9] S. Graham, S. Quirke-McFarlane, V. Auyeung, and J. Weinman, "Interventions Designed to Improve Adherence to Growth Hormone Treatment for Pediatric Patients and Their Families: A Narrative Review." Pharmaceutics, vol. 14, pp. 2373, Nov. 2022, doi: 10.3390/pharmaceutics14112373.

[10] P. H. Su et al., "Investigating the Impact of the TUITEK® Patient Support Programme, Designed to Support Caregivers of Children Prescribed Recombinant Human Growth Hormone Treatment in Taiwan." Front. Endocrinol. (Lausanne), vol. 13, pp. 897956, May 2022, doi: 10.3389/fendo.2022.897956.

[11] A. Halabi, B. Martin, E. Koledova, G. Giunti, and P. Dimitri, "A patient-centric approach to connected health solutions in paediatric growth hormone therapy." In: ESPE Abstracts, Bioscientifica, 2022, Available from: https://abstracts.eurospe.org/hrp/0095/hrp0095p1-500.

[12] M. Rosenlund, U. M. Kinnunen, and K. Saranto, "The Use of Digital Health Services Among Patients and Citizens Living at Home: Scoping Review." J Med Internet Res, vol. 25, pp. e44711, Mar. 2023, doi: 10.2196/44711.

[13] J. Kesavadev, A. Shankar, P. B. S. Pillai, G. Krishnan, and S. Jothydev, "Cost-Effective Use of Telemedicine and Self-Monitoring of Blood Glucose via Diabetes Tele Management System (DTMS) to Achieve Target Glycosylated Hemoglobin Values Without Serious Symptomatic Hypoglycemia in 1,000 Subjects with Type 2 Diabetes Mellitus—A Retrospective Study." Diabetes Technology & Therapeutics, vol. 14, pp. 772-776, Sep. 2012, doi: 10.1089/dia.2012.0088.

[14] V. Tornincasa et al., "Integrated Digital Health Solutions in the Management of Growth Disorders in Pediatric Patients Receiving Growth Hormone Therapy: A Retrospective Analysis." Front. Endocrinol., vol. 13, pp. 882192, Jun. 2022, doi: 10.3389/fendo.2022.882192.

[15] E. Koledova, V. Tornincasa, and P. van Dommelen,"Analysis of real-world data on growth hormone therapy adherence using a connected injection device." BMC Med. Inform. Decis. Mak., vol. 20, pp. 176, Jul. 2020, doi: 10.1186/s12911-020-01183-1.

[16] P. Dimitri and M. O. Savage, "Artificial intelligence in paediatric endocrinology: conflict or cooperation." J. Pediatr. Endocrinol. Metab., vol. 37, pp. 209-221, Mar. 2024, doi: 10.1515/jpem-2023-0554.

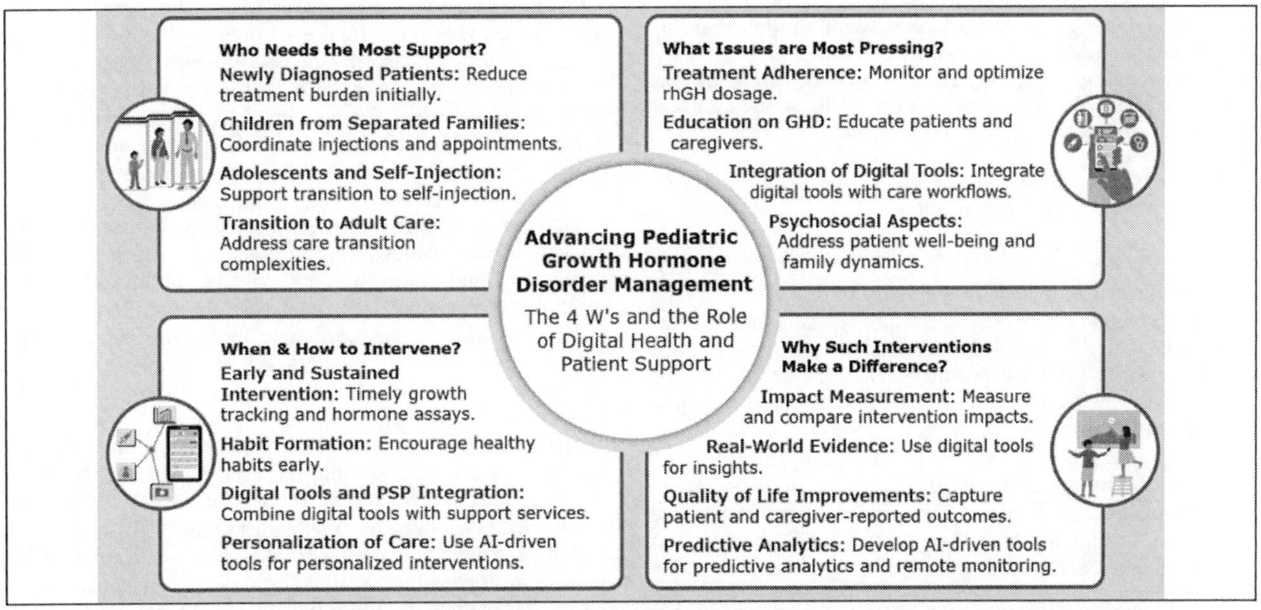


Figure 1. The 4 W's and the role of digital health technologies and patient support.

# Zero-Shot Super-Resolution for Low-Dose CBCT Images Using Lightweight StereoMamba

Simin Mirzaei, Zhenchao Ma, Hamid Reza Tohidypour, and Panos Nasiopoulos
*Electrical & Computer Engineering, University of British Columbia*
Vancouver, BC, Canada
Email: {siminmirzaei, zhenchaoma, htohidyp, panosn}@ece.ubc.ca

***Abstract*—Cone-Beam Computed Tomography (CBCT) is a crucial imaging tool in medical diagnostics, but low-dose scans—necessary for minimizing patient radiation exposure—often suffer from degraded spatial resolution. Enhancing the visual quality of these scans is essential for accurate diagnosis and treatment planning. This paper introduces two primary contributions to address this challenge. First, we systematically evaluate the effectiveness of state-of-the-art super-resolution techniques on low-dose CBCT images. Due to the scarcity of real CBCT datasets, which are limited by radiation exposure constraints, we explore the potential of pre-trained stereo super-resolution models originally developed for RGB images. Unlike traditional CBCT datasets that rely on artificially synthesized image pairs, we employ a zero-shot approach to assess the adaptability of these pre-trained models to CBCT imaging. Second, our analysis reveals that existing deep-learning-based super-resolution networks struggle to generalize effectively to CBCT data. To address this, we develop a lightweight adaptation of StereoMamba, a model optimized for natural images, and tailor it to the structural characteristics of CBCT scans. Without requiring additional training, our optimized network achieves the highest Peak Signal-to-Noise Ratio (PSNR) and Structural Similarity Index Measure (SSIM) scores among all tested models, significantly enhancing CBCT image quality. Our approach makes a significant contribution to improving the quality of low-dose CBCT imaging and provides a path forward for improving diagnostic accuracy and clinical outcomes without increasing radiation risk.**

***Keywords*—Cone-Beam Computed Tomography; spatial resolution; StereoMamba; lightweight models; zero-shot learning.**

## I. INTRODUCTION

Cone-Beam Computed Tomography (CBCT) is a crucial imaging technique extensively utilized in dentistry and medical diagnostics, offering detailed 3D visualization of anatomical structures [1][2]. In dental practice, CBCT has revolutionized diagnosis and treatment planning by providing accurate imaging of the maxillofacial region, enabling clinicians to easily examine axial, sagittal, coronal, and custom plane sections [3]. However, achieving high-resolution imaging necessitates increased radiation exposure, heightening the risk of radiation-induced cancers [4]. To mitigate this issue, low-dose CBCT imaging has been developed, which leads to fewer X-ray photons reaching the detector, resulting in lower Signal-to-Noise Ratio (SNR) and reduced spatial resolution [5][6].

Numerous studies have investigated super-resolution, primarily using Single-Image Super-Resolution (SISR) techniques, across various applications. To some extent, these methods can also be applied to CBCT imaging [7][8][9][10][11]. For instance, Hwang et al. [7] investigated the use of a Very Deep Super-Resolution (VDSR) network to restore compressed CBCT images, demonstrating a significant improvement in image quality. Their findings highlight the potential of this approach for clinical applications, enabling reduced storage requirements while maintaining diagnostic accuracy. In another study, Oyama et al. [8] presented a method to suppress artifacts in CBCT images by combining low-resolution images with corresponding high-resolution ones through super-resolution techniques, thereby enhancing clarity and reducing artifacts. Liu et al. [9] uses a Generative Adversarial Network (GAN) to enhance the resolution of low-dose CT images by employing a pyramidal attention model and multiple residual dense blocks to focus on high-frequency image information. Shen et al. [10] proposed a new super-resolution network that uses deep gradient information to guide the reconstruction of CT images, merging gradient image features into the super-resolution branch to enhance structural preservation and detail restoration, ultimately improving image quality. Saharia et al. [11] introduced SR3, a method that uses denoising diffusion probabilistic models to enhance image resolution. Moreover, Liang et al. [12] proposed a hierarchical transformer-based framework for image super-resolution that utilizes the Swin Transformer's efficient window-based self-attention mechanism to effectively capture both local and global dependencies, achieving state-of-the-art performance and surpassing traditional Convolutional Neural Network (CNN)-based methods in both quality across diverse benchmarks. These existing techniques primarily rely on single-image information to enhance the resolution of individual images.

Recognizing this limitation, research on image super-resolution has expanded beyond single-image techniques, with numerous studies exploring approaches that incorporate multiple dependent input images, such as Stereo Super-Resolution (SSR) techniques. These approaches leverage shared information between the input images, resulting in higher spatial resolution for each image [13][14][15][16][17][18]. For example, Lu et al. [13] proposed a two-stage Cycle-consistency network which reconstructs thin-slice MR images by leveraging information from adjacent thick slices to enhance resolution and ensure consistency. In another study, Zhao et al. [14] proposed a self-supervised deep learning algorithm for MRI applications that utilizes shared data between adjacent slices to enhance MRI resolution. By learning spatial correlations across neighboring slices, this method improves both resolution and image quality. Sood et al. [15] presented a method for accurately aligning pre-surgical prostate MRI with histopathology images using super-resolution volume reconstruction. They developed a multi-image super-resolution GAN that leverages multiple lower-resolution MRI images as input to generate high-resolution 3D MRI volumes. Cai et al. [16] proposed a deep learning framework that enhances the segmentation of hepatic ducts in

CT scans. To achieve high-quality segmentation, their method includes an inter-slice super-resolution subnetwork which uses information from neighboring slices in a CT volume to generate intermediate slices, effectively increasing the resolution by interpolating between the existing CT slices. Stereo image super-resolution, which focuses on reconstructing high-resolution details from low-resolution left and right image pairs, has gained significant attention in recent years [17][18]. For example, Chu et al. [17] proposed a novel approach for stereo image super-resolution by extending the NAFNet architecture with components that effectively leverage cross-view information between left and right images. This design enhances spatial coherence and depth consistency, achieving state-of-the-art performance on stereo image super-resolution benchmarks with superior visual quality compared to single-image super-resolution methods. Recently, Ma et al. [18] proposed StereoMamba, a stereo image super-resolution method built on Structured State Space Models (SSMs). The method leverages the Mamba architecture to effectively capture inter-view correlations between left and right image pairs and a Stereo Bi-directional Cross-Attention Module (SBCAM) to further enhance stereo view coherence. Experimental results showed that StereoMamba outperforms state-of-the-art methods for both single and stereo image super-resolution tasks.

In this paper, we present two key contributions. First, we evaluate the performance of state-of-the-art super-resolution methods on low-dose CBCT scans to assess their effectiveness for CBCT applications. Given the scarcity of real CBCT datasets due to radiation exposure constraints, we focus on leveraging pre-trained deep learning models for stereo super-resolution. Since existing CBCT datasets primarily consist of artificially generated pairs with limited practical utility, we adopt a zero-shot strategy, utilizing models pre-trained on large RGB datasets to examine their applicability and generalizability to CBCT imaging.

Second, based on our findings that complex pre-trained super-resolution networks struggle to generalize effectively to CBCT images, we develop a lightweight version of StereoMamba, the network shown to excel on natural images. By adapting this network to the simpler structural patterns of CBCT images, our lightweight model performs exceptionally well on CBCT scans without requiring retraining. This model outperforms all other state-of-the-art super-resolution networks tested, achieving the highest Peak Signal-to-Noise Ratio (PSNR) and Structural Similarity Index Measure (SSIM) scores, making a significant contribution to improving the quality of low-dose CBCT imaging and providing a path forward for improving diagnostic accuracy and clinical outcomes without increasing radiation risk.

The rest of this paper is organized as follows: Section II presents our objective evaluation of state-of-the-art SISR and SSR methods on CBCT images. Section III details the architecture of our proposed lightweight StereoMamba network, its performance evaluation and discusses the results. Finally, Section V concludes the paper.

## II. Performance Evaluation of Super-Resolution Methods for Low-Dose CBCT Scans

In low-dose CBCT imaging, reducing radiation exposure results in lower-resolution scans, making it more challenging for dentists and medical professionals to accurately diagnose conditions. The decrease in X-ray photons reaching the detector lowers the SNR, increases noise, and obscures fine anatomical details, ultimately reducing spatial resolution [6]. This limitation is particularly problematic in dental applications, where high-resolution imaging is essential for precise diagnosis and treatment planning.

We conduct a comprehensive evaluation of super-resolution methods, comparing both SISR and SSR approaches to assess their effectiveness for CBCT imaging. It is worth noting that in general while SISR models perform well on single images, they are inherently limited when compared to SSR methods, as they fail to utilize the redundant spatial information present in adjacent scans. In contrast, SSR methods leverage cross-view correlations, making them more effective for tasks involving medical imaging modalities, such as CBCT. Despite the limitations of SISR approaches, we decided to include the SwinIR network [12] in our evaluation, as it represents the most advanced SISR method available.

Theoretically, deep learning-based super-resolution approaches can enhance low-dose CBCT scans, but their effectiveness is hindered by the scarcity of high-quality training datasets. Unlike natural image super-resolution, which benefits from large datasets of paired low- and high-resolution images, CBCT imaging faces significant constraints. Acquiring both low-dose and high-dose scans from the same patient is not only impractical but also ethically impossible due to radiation exposure risks. As a result, existing CBCT datasets are small and often rely on synthetically generated high-resolution images, which may introduce biases and fail to accurately represent real-world conditions. This data limitation makes it difficult for deep learning models to generalize effectively, often leading to overfitting and reducing their practical applicability for improving CBCT image quality.

To overcome these challenges, we avoid retraining state-of-the-art super-resolution networks used in our evaluation and instead we use the corresponding pre-trained deep learning models. For the case of stereo super-resolution models, given the structural and spatial correlation between adjacent CBCT slices, we treat them as stereo image pairs, allowing us to exploit inter-slice redundancies to enhance image quality while minimizing radiation exposure. Figure 1 illustrates a sequence of transverse slices from a CBCT volume, highlighting the structural continuity between adjacent scans.

For our evaluation of SSR approaches, we selected the top-performing models, specifically variants of NAFSSR [17], along with StereoMamba [18]. The former include NAFSSR-T, the smallest version with minimal parameters; NAFSSR-S, an optimized variant balancing efficiency and performance; NAFSSR-B, a balanced model that maintains high reconstruction quality while reducing computational complexity; and NAFSSR-L, the largest variant designed for maximal reconstruction fidelity. Additionally, we included StereoMamba, which has been shown to outperform all NAFSSR variants on natural images.

By adopting this zero-shot learning strategy, we assess the generalizability of both SISR and SSR pre-trained models to CBCT imaging without additional fine-tuning, demonstrating their potential to improve resolution despite being trained on natural images.

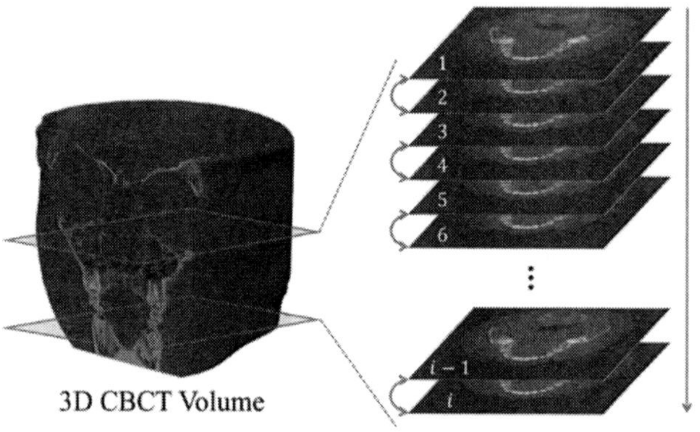


Figure 1. A series of 2D slices from the transverse view in a 3D CBCT volume.

### A. *Results of Our Comparative Evaluation*

We use single CBCT slices as input for the SISR model (SwinIR) and adjacent CBCT slices as input pairs for SSR models (NAFSSR and StereoMamba), treating them as two-view images to capture inter-slice dependencies. The evaluation results are summarized in Table I. We observe that all tested networks achieve similar PSNR and SSIM values, with StereoMamba performing the worst. This is unexpected, as StereoMamba has demonstrated superior performance on natural images [18]. These results suggest that although StereoMamba has the most promising architecture, it does not generalize well to CBCT images. This anomaly is due to the fact that grayscale CBCT images tend to have simpler texture and structures and lack color channels, forcing larger, more complex models to overfit to dataset-specific details, amplifying noise or irrelevant features and resulting in suboptimal performance.

Motivated by this finding, we decided to develop a lightweight version of the StereoMamba network, specifically designed to better adapt to the characteristics of CBCT images. The details of this optimized model and its performance are presented in the following section.

## III. Overall Architecture of Our Lightweight StereoMamba and Performance Evaluation

In this contribution, we modified the original StereoMamba network presented in [18] to a lightweight version, as illustrated in Figure 2. The original StereoMamba model, while effective for natural images, struggled to generalize to grayscale CBCT images due to their simpler textures, fewer structural variations,

TABLE I. Performance Comparison of State-of-the-Art SISR and SSR Methods for CBCT Resolution Enhancement (#S and #Params Indicate the Scale Factor and the Number of Parameters, Respectively).

| Method | #S | #Params | PSNR | SSIM |
|---|---|---|---|---|
| SwinIR | x2 | 11.28M | 36.32 | 0.9849 |
| NAFSSR-T | x2 | 0.45M | 36.72 | 0.9865 |
| NAFSSR-S | x2 | 1.54M | 36.65 | 0.9859 |
| NAFSSR-B | x2 | 6.77M | 36.47 | 0.9857 |
| NAFSSR-L | x2 | 23.79M | 36.08 | 0.9846 |
| StereoMamba | x2 | 7.55M | 35.88 | 0.9840 |

and lack of color channels. To enhance generalization, we reduced the number of Residual State Space Groups (RSSGs) to four and the number of SBCAMs to three, as shown in Figure 2. Key differences between our lightweight StereoMamba and the original model include a reduction in the embedding dimension—from 120 to 60—in both the shallow and deep feature extraction stages, as well as a decrease in the expansion ratio of the Vision State Space Module (VSSM) from 2 to 1.2 [18][19]. The process starts with two consecutive low-resolution CBCT slices, labeled as slices $i-1$ and $i$, serving as inputs. Each pair of neighboring slices is processed sequentially, ensuring no overlap between the groups. The slices first pass through a $3 \times 3$ convolutional layer to extract fundamental features, capturing key patterns necessary for subsequent processing. The extracted features are then fed into the deep feature extraction section, which employs four RSSGs connected through residual links. These modules process input features through linear transformations, depth-wise convolutions, and SiLU activation to capture both local and global spatial correlations. Residual connections are included to preserve input information and improve feature propagation. This design ensures feature continuity, enhances gradient flow during training, and refines feature representations. To strengthen the interaction between adjacent slices $i-1$ and $i$, the deep feature extraction section incorporates SBCAM modules. These modules employ bi-directional cross-attention mechanisms to enable a synergistic exchange of contextual information between adjacent CBCT slices, improving the network's ability to capture inter-slice relationships. Finally, the reconstruction module processes the refined feature maps from the deep feature extraction section, integrating them to generate high-resolution reconstructions of input adjacent CBCT slices.

Our proposed complexity reduction allows this lightweight version to avoid memorizing dataset-specific noise and instead to focus on essential structural details, leading to more robust feature extraction. Additionally, fewer layers reduce the accumulation of redundant or noisy representations, mitigating the risk of amplifying artifacts. In summary, this streamlined architecture aims to enhance generalization without retraining and achieve high-quality super-resolution results, making it well-suited for CBCT applications.

Table II shows the performance of our pretrained lightweight StereoMamba model, as well as that of the state-of-the-art networks included in Table I in terms of PSNR and SSIM. The evaluation results demonstrate that our lightweight StereoMamba model outperforms all other methods, achieving superior image quality in terms of both structural preservation and perceptual similarity. Notably, NAFSSR-T, which also has a relatively small number of parameters, also performs well, reinforcing the idea that lighter models are more robust to domain shifts from RGB to grayscale CBCT images. These models avoid overfitting by focusing on fundamental spatial features rather than dataset-specific variations. In summary, our evaluation demonstrates that our lightweight StereoMamba model effectively balances complexity and performance, achieving state-of-the-art results in CBCT image super-resolution while maintaining computational efficiency. Its ability to generalize across different datasets, combined with reduced inference time, makes it a strong candidate for practical applications in dental and medical imaging.

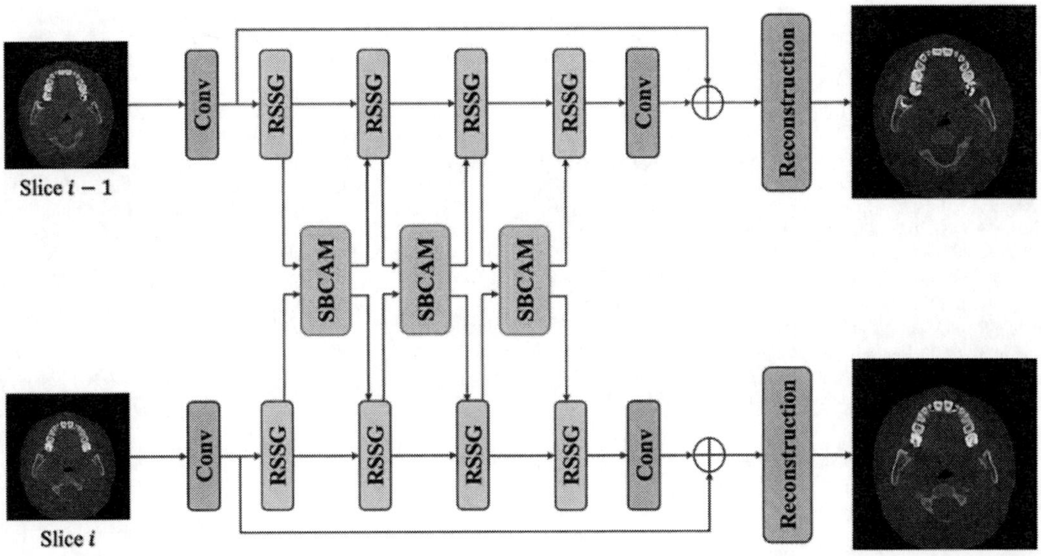


Figure 2. Architecture of our lightweight StereoMamba network for enhancing the resolution of adjacent slices in a 3D CBCT volume.

TABLE II. PERFORMANCE COMPARISON OF OUR LIGHTWEIGHT STEREOMAMBA AGAINST THE STATE-OF-THE-ART SUPER-RESOLUTION METHODS (#S AND #PARAMS INDICATE THE SCALE FACTOR AND THE NUMBER OF PARAMETERS, RESPECTIVELY).

| Method | #S | #Params | PSNR | SSIM |
|---|---|---|---|---|
| SwinIR | x2 | 11.28M | 36.32 | 0.9849 |
| NAFSSR-T | x2 | 0.45M | 36.72 | 0.9865 |
| NAFSSR-S | x2 | 1.54M | 36.65 | 0.9859 |
| NAFSSR-B | x2 | 6.77M | 36.47 | 0.9857 |
| NAFSSR-L | x2 | 23.79M | 36.08 | 0.9846 |
| StereoMamba | x2 | 7.55M | 35.88 | 0.9840 |
| **StereoMamba-Light** | x2 | **0.9M** | **38.03** | **0.9886** |

## IV. CONCLUSION AND FUTURE WORK

In this paper, we conducted a comprehensive evaluation of super-resolution techniques for low-dose CBCT imaging, comparing state-of-the-art SISR and SSR methods. A key challenge in applying deep learning-based super-resolution to CBCT imaging is the scarcity of high-quality paired training datasets. To mitigate this limitation, we adopted a zero-shot learning approach, utilizing pre-trained networks without additional fine-tuning. Our evaluation demonstrated that the original StereoMamba model, despite its success in natural image super-resolution, struggled to generalize to grayscale CBCT images. Motivated by this finding, we developed a lightweight version of StereoMamba, reducing model complexity while preserving essential spatial features. The results indicate that our proposed lightweight StereoMamba model outperforms existing SISR and SSR methods, achieving improved spatial resolution and structural preservation in low-dose CBCT scans. By reducing network complexity and focusing on fundamental spatial correlations, our lightweight StereoMamba provides a practical and effective solution for enhancing CBCT image quality without requiring extensive retraining. This advancement has significant implications for dental and medical applications, where improved CBCT resolution can enhance diagnostic accuracy while minimizing radiation exposure. Future work will explore domain-specific fine-tuning strategies and further optimization of SSR architectures to enhance generalizability across different medical imaging modalities.

## REFERENCES

[1] B. Schlueter, K. B. Kim, D. Oliver, and G. Sortiropoulos, "Cone beam computed tomography 3D reconstruction of the mandibular condyle," The Angle Orthodontist, vol. 78, no. 5, pp. 880–888, 2008.

[2] W. De Vos, J. Casselman, and G. R. J. Swennen, "Cone-beam computerized tomography (CBCT) imaging of the oral and maxillofacial region: A systematic review of the literature," International journal of oral and maxillofacial surgery, vol. 38, no. 6, pp. 609–625, 2009.

[3] C. Angelopoulos, "Anatomy of the maxillofacial region in the three planes of section," Dental Clinics, vol. 58, no. 3, pp. 497-521, 2014.

[4] J. B. Carter, J. D. Stone, R. S. Clark, and J. E. Mercer, "Applications of cone-beam computed tomography in oral and maxillofacial surgery: an overview of published indications and clinical usage in united states academic centers and oral and maxillofacial surgery practices," Journal of Oral and Maxillofacial Surgery, vol. 74, no. 4, pp. 668–679, 2016.

[5] S. Kawauchi, K. Chida, Y. Hamada, and W. Tsuruta, "Image quality and radiation dose of conventional and wide-field high-resolution cone-beam computed tomography for cerebral angiography: a phantom study," Tomography, vol. 9, no. 5, pp. 1683-1693, 2023.

[6] L. Yu et al., "Dose and image quality evaluation of a dedicated cone-beam CT system for high-contrast neurologic applications," American Journal of Roentgenology, vol. 194, no. 2, pp. W193-W201, 2010.

[7] J. J. Hwang, Y. H. Jung, B. H. Cho, and M. S. Heo, "Very deep super-resolution for efficient cone-beam computed tomographic image restoration," Imaging Science in Dentistry, vol. 50, no. 4, p. 331, 2020.

[8] A. Oyama et al., "Image quality improvement in cone-beam CT using the super-resolution technique," Journal of radiation research, vol. 59, no. 4, pp. 501-510, 2018.

[9] X. Liu et al., "Super-resolution reconstruction of ct images based on multi-scale information fused generative adversarial networks," Annals of Biomedical Engineering, vol. 52, no. 1, pp. 57-70, 2024.

[10] Y. Shen et al., "Ct image super-resolution under the guidance of deep gradient information," Journal of X-Ray Science and Technology, vol. 33, no. 1, pp. 58-71, 2025.

[11] C. Saharia et al., "Image super-resolution via iterative refinement," IEEE transactions on pattern analysis and machine intelligence, vol. 45, no. 4, pp. 4713-4726, 2022.

[12] J. Liang et al., "Swinir: Image restoration using swin transformer," Proceedings of the IEEE/CVF international conference on computer vision, pp. 1833-1844, 2021.

[13] Z. Lu, Z. Li, J. Wang, J. Shi, and D. Shen, "Two-stage self-supervised cycle-consistency network for reconstruction of thin-slice MR images," Medical Image Computing and Computer Assisted Intervention–MICCAI 2021: 24th International Conference, Strasbourg, France, pp. 3-12, 2021.

[14] C. Zhao et al., "SMORE: a self-supervised anti-aliasing and super-resolution algorithm for MRI using deep learning," IEEE transactions on medical imaging, vol. 40, no. 3, pp. 805-817, 2020.

[15] R. R. Sood et al., "3D Registration of pre-surgical prostate MRI and histopathology images via super-resolution volume reconstruction," Medical image analysis, vol. 69, p. 101957, 2021.

[16] N. Cai et al., "W-Shaped Net: An Inter-Slice Super-Resolution Segmentation Deep Network for CT Scans of Hepatic Ducts," Electronics, vol. 14, no. 2, p. 321, 2025.

[17] X. Chu, L. Chen, and W. Yu, "Nafssr: Stereo image super-resolution using nafnet," Proceedings of the IEEE/CVF conference on computer vision and pattern recognition, pp. 1239-1248, 2022.

[18] Z. Ma, H. R. Tohidypour, P. Nasiopoulos, and V. C. M. Leung, "StereoMamba: Enhancing Stereo Image Super-Resolution with Structured State Space Models and Bi-Directional Cross Attention," ICASSP 2025-2025 IEEE International Conference on Acoustics, Speech and Signal Processing (ICASSP), pp. 1-5, 2025.

[19] H. Guo et al., "Mambair: A simple baseline for image restoration with state-space model," European conference on computer vision, pp. 222-241, 2024.